T0305569

Root Cause Analysis (RCA) for the Improvement of Healthcare Systems and Patient Safety

Root Cause Analysis (RCA) for the Improvement of Healthcare Systems and Patient Safety

David Allison, CPPS
Harold Peters, P.Eng.

CRC Press
Taylor & Francis Group
Boca Raton   London   New York

CRC Press is an imprint of the
Taylor & Francis Group, an **informa** business

First edition published 2022
by CRC Press
6000 Broken Sound Parkway NW, Suite 300,
Boca Raton, FL 33487–2742

and by CRC Press
2 Park Square, Milton Park, Abingdon, Oxon, OX14 4RN

CRC Press is an imprint of Taylor & Francis Group, LLC

Library of Congress Cataloging-in-Publication Data
Names: Allison, David, author. | Peters, Harold, author.
Title: Root cause analysis (RCA) for the improvement of healthcare systems and patient safety / David Allison, CPPS and Harold Peters, P.Eng.
Description: Boca Raton : CRC Press, 2021. | Includes bibliographical references and index.
Identifiers: LCCN 2021009377 (print) | LCCN 2021009378 (ebook) |
   ISBN 9781032035925 (hardback) | ISBN 9781032036014 (paperback) |
   ISBN 9781003188162 (ebook)
Subjects: LCSH: Medical errors—Prevention. | Medical care—Safety measures. |
   Root cause analysis.
Classification: LCC R729.8 .A447 2021 (print) | LCC R729.8 (ebook) |
   DDC 610.28/9—dc23
LC record available at https://lccn.loc.gov/2021009377
LC ebook record available at https://lccn.loc.gov/2021009378

ISBN: 978-1-032-03592-5 (hbk)
ISBN: 978-1-032-03601-4 (pbk)
ISBN: 978-1-003-18816-2 (ebk)

Typeset in Times
by Apex CoVantage, LLC

Contents

PART I Building an Understanding of RCA

PART II   Root Cause Analysis Champions

Preface

This is a book about root cause analysis (RCA) in the context of healthcare and is intended for healthcare professionals leading the analysis of patient harm events, those executives and leaders directing and managing patient safety, and those advocating for evaluation of patient harm using best practices. The intent is to equip healthcare caregivers[1] at all levels to employ this powerful tool for patient safety. Why is this important? Because patients continue to be unintentionally harmed as the result of system and human failures. Preventing harm requires healthcare organizations to learn from unintended events that result in harm and allow those learnings to define the necessary actions to eliminate the causes of harm. Substantial opportunities also exist in learning from close call events and taking action to eliminate those causes as well. Both cases require reliably finding the sources of failure and implementing permanent preventive actions.

This book is the outgrowth of the authors' 15 years of learning, facilitating, and teaching RCA in healthcare. David has participated in and led RCAs as the manager of a hospital psychiatric unit and as a risk management and patient safety professional. Harold has supported RCAs as a healthcare improvement engineer. *Root Cause Analysis for the Improvement of Healthcare Systems and Patient Safety* is based on the curriculum we developed for teaching RCA facilitation. Our goal is to provide the necessary knowledge, combined with practical application, to allow the reader to embark on leading successful RCAs.

Part I begins with building an understanding of RCA. Chapter 1 examines the need for RCA. Chapter 2 recognizes RCA as one tool among many and addresses the question of when to employ an RCA. Chapter 3 describes the pre-work for an RCA team meeting. Chapter 4 introduces the Logic Tree RCA methodology.[2] Chapter 5 suggests how to construct effective action plans. Chapter 6 focuses on practical aspects of facilitation for an RCA team meeting.

Having done a deep dive into understanding and facilitating RCA in Part I, Part II is written to support the champions of RCA within your own healthcare setting. Chapter 7 goes deeper into the roles for successful RCA. Chapter 8 looks at common ways RCAs fail and provides countermeasures to the barriers faced. Chapter 9 examines trending and how RCA fits in a larger strategy to achieve the goal of zero harm. Chapter 10 presents the curriculum outline for training.

NOTES

1. We use the term "caregivers" for those involved in the healthcare enterprise, including employed and allied providers and clinicians, as well as those who work directly, or indirectly, in patient care.
2. The Logic Tree is based on the PROACT model by Reliability Center, Inc. We are grateful to Robert Latino, RCI Chief Executive Officer, who has been instrumental in our understanding RCA and a proponent of RCA for improving healthcare.

Acknowledgments

The path to this book began in 2005 with discussing how we could better investigate unintended events and make healthcare safer. The exploration and dialogue that followed continued over the years, culminating in agreement over dinner to collaborate and turn our shared learnings into book form. Our professional collaboration and all that entails, support and respectful challenge of one another, have enriched our practices. It has taken courage to learn from our own mistakes, and perseverance to continually improve. We are better for it.

We are grateful for the many people who have supported this exploration, beginning with Pam Bristol, Elaine Dunda, and other Risk Management and Quality leaders willing to standardize our practice of RCA on a "new" methodology. Bob Latino and his colleagues at Reliability Center Inc. have been instrumental in helping us understand, apply, and find rigor in this "new" form of RCA. Bob is a staunch champion for patient safety and a leader in adapting decades of industrial root cause experience and expertise to healthcare. His advice and encouragement in writing the book was influential. We recognize the physician leaders who have supported our efforts and provided ongoing improvement feedback, including Dr. Stephanie Jackson, Dr. Andrea Halliday, Dr. Lawrence Neville, and Dr. Sudhakar Karlapudi. We received encouragement from Natalie Johnston, Laura Walker, Kerri Ronayne, Carina Oei, Lisa Wolfe, Janet Burrows, Jackie Jones-Baily, and Ashley Groshong: true champions of RCA. Special thanks go to Molly Rank and Jaime Leon, who have partnered in teaching RCA, improved the curriculum, and coached RCA facilitation. We appreciate their reviews of drafts and instructive comments. To all of you with whom we have shared the journey of patient safety improvement, this book would not have been possible without your trust, patience, and commitment.

Our thanks to Erin Harris and Cindy Renee Carelli at CRC Press / Taylor & Francis Group for guiding us on this journey as new authors.

Thank you doesn't begin to express the appreciation for the support of family: Braden, Camden, and Katie. And, especially, gratitude to our spouses: Leslie Hall, who read and edited many drafts, and Dr. Juanita Doerksen, both of whom were understanding of the long hours spent at the computer on this project.

Authors

David Allison, CPPS, has 15+ years of facilitating RCA teams and teaching RCA methodology for patient safety and risk management professionals. He has over 30 years of experience in healthcare and has provided leadership in behavioral health, risk management, and patient safety settings. David has been the process owner for the safety value stream across a healthcare system, helping to reduce the rate of serious safety events with tools such as RCA.

Harold Peters, P.Eng., is an improvement professional with extensive experience in healthcare, service, government, and manufacturing. During his 15+ years in healthcare, he led Lean project and transformation work, facilitated RCAs, and introduced other improvement methodologies like Work Simplification, Theory of Constraints, and Operations Research. In system leadership roles, he established and led the process improvement strategy, structure, standards, and resources for two large healthcare systems across multiple states, and led the system patient safety department in one of the organizations, developing strategy, structure, standards, and teaching RCA methodologies.

Part I

Building an Understanding of RCA

1 The Need for Root Cause Analysis (RCA)

1.1 WHY INVESTIGATE HEALTHCARE UNINTENDED EVENTS?

Let us begin with a reflection on why we should devote the time and resources to thorough and rigorous root cause analysis (RCA). Empathy is an oft-strived-for attribute in healthcare and frequently associated with compassion. In fact, the Latin root of compassion means *to suffer alongside*. Being empathic as caregivers helps us appreciate the patients' perspective: we are better attuned to their fears and concerns, hopes, and desires. Such compassion yields greater alignment of goals for recovery. When unintended events occur, resulting in patients being harmed, our empathy motivates a caring response. However, compassion as empathy alone will not be sufficient to prevent the unintended event from reoccurring, resulting in the harm of a future patient.

Compassion, comprised of both empathy and intellectual understanding, is called for in caring for patients who have been harmed and for those who have made errors that result in harm. In addition to our affective response, compassion in healthcare requires using our intellect. We need to employ the scientific method to discover causes and learn how to prevent harm. Don Berwick[1] drew lessons for the future of healthcare from Norman MacLean's book, *Young Men and Fire*.[2] In the moving story of the 1949 Mann Gulch fire near Helena, Montana, in which 13 young men lost their lives, MacLean carefully reconstructs the fateful sequence of events. MacLean explains how the factors of weather, terrain, and characteristics of fire all combined to result in a small fire blowing up into a conflagration, and he helps understand the decisions made by members of the crew that resulted in their lack of escape. He also tells the remarkable story of crew chief Wag Dodge, who spontaneously invented the escape fire and survived. The power of root cause analysis is in the depth of understanding gained in breaking down how a tragic event occurred, to prevent it from reoccurring. That is why we perform RCA on healthcare unintended events.

1.2 CARING FOR CAREGIVERS

We are in a period of transition from a history of "shame and blame" to one of "caring for the caregivers." Physicians, nurses, technologists, pharmacists, and therapists are among the disciplines in healthcare most vulnerable to making human errors. As an industry, we have been constrained under a system of litigation that drives practitioners to be advised not to speak about an error with anyone but their legal representatives. This milieu of shame and silence can compound tragedy upon tragedy.

European researchers[3] found healthcare professionals identified three needs after an error caused patient harm. There is a need for peer support. There is a desire for support from management. And there is a need for stable and transparent analysis. Models of peer support are emerging across healthcare. Sue Scott[4] of the University of Missouri has written extensively on the impact of a peer support program. Others such as the Medically Induced Trauma Support Services (MITSS)[5] and the Johns Hopkin Resilience in Stressful Events (RISE)[6] programs serve as models for emotionally supporting caregivers. Matthew Syed[7] suggests the need for resilience: "The capacity to face up to failure, and to learn from it. Ultimately, that is what growth is all about."

Thorough and rigorous RCA is one way in which a healthcare organization supports a shift from blaming caregivers to becoming a learning organization. A premise for our RCA practice has been the assumption that those involved in unintended events are skilled and dedicated caregivers: it has been rare to find reckless behavior. However, we also understand that the same error will likely reoccur unless there are changes made to prevent it.

We advocate for including caregivers involved in an unintended event on the RCA team. Others have suggested a different approach out of concern "that they may feel guilty and insist on corrective measures that are above and beyond what is prudent, or they may steer the team away from their role in the event and activities that contributed to the event."[8] Our approach to RCA team composition will be spelled out in Chapter 3. For now, we note one of the important reasons for doing RCA is the opportunity it affords caregivers to be involved in improving safety for future patients and their colleagues. Gaining the understanding of how the system failed and how the caregiver is not a terrible person is a consistently positive experience for those directly involved in a healthcare error.

RCA that rigorously drives to the latent system vulnerabilities, as commonly depicted by James Reason's "Swiss cheese" model,[9] aligns with an understanding of accountability based on just culture. Caregivers are more likely to speak up and identify safety concerns in an atmosphere where they feel psychologically safe. Root cause analysis helps to understand the mutual accountabilities of the individual caregiver and the healthcare organization through understanding the process and system-related failures that result in healthcare-unintended events.

1.3 SYSTEMS APPROACH

The National Patient Safety Foundation looked back 15 years after the publication in 1999 of the seminal work, *To Err Is Human*.[10] The findings were published in the white paper, "Free from Harm: Accelerating Patient Safety Improvement Fifteen Years after *To Err Is Human*."[11] The panel observed that two general approaches had evolved to improving patient safety over the prior 15 years. One strategy was to identify specific safety problems and approach them on a project-by-project basis. The panel found that this strategy had not led to widespread improvements despite remarkable singular successes. A second strategy sought to emulate other industries that took more of a systems approach. This approach led to the following conclusion, "By taking into account systems design, human failures, human factors engineering, safety culture, and error reporting and analysis, the systems approach epitomizes a more comprehensive view."[12] To understand how RCA contributes to a systems approach, and its value for improving patient safety, it is helpful to reflect on the experience of commercial aviation safety.

1.3.1 AVIATION'S BREAKTHROUGH

The commercial airline industry in the United States, as early as the 1970s, had achieved significant progress in preventing crashes and fatalities. Jet aircraft, with increased capabilities to avoid hazards, had become the industry norm. Human factors engineering had aided cockpit design. Checklists had been utilized since World War II. But even with these improvements, crashes and fatalities continued. More recently, during the years 2010–2017, U.S. commercial carriers achieved a record of zero fatalities (see Table 1.1 for the data).[13] How did they accomplish this? The story begins with the crash of United Airlines Flight 173 on December 28, 1978.[14]

TABLE 1.1
U.S. Commercial Carrier Passenger Fatalities

Passenger Injuries and Injury Rates, 1999 through 2018, for U.S. Air Carriers Operating under 14 CFR 121

Year	Passenger Injuries		Passenger Enplanements (millions)	Million Passenger Enplanements per Passenger Fatality
	Fatalities	Serious Injuries		
1999	10	46	676	67.6
2000	83	13	701	8.4
2001	483	7	629	2.5
2002	0	11	619	No Fatalities
2003	19	10	654	34.4
2004	11	3	711	64.6
2005	18	2	743	41.3
2006	47	4	747	15.9
2007	0	3	770	No Fatalities
2008	0	6	745	No Fatalities
2009	45	14	706	15.7
2010	0	5	723	No Fatalities
2011	0	4	734	No Fatalities
2012	0	3	740	No Fatalities
2013	0	1	746	No Fatalities
2014	0	0	766	No Fatalities
2015	0	8	801	No Fatalities
2016	0	4	826	No Fatalities
2017	0	1	851	No Fatalities
2018	1	10	891	890.9

Flight 173 crashed after running out of fuel while the pilot and cockpit crew were trying to determine if the landing gear was locked down upon approach to the Portland International Airport. They did not hear the expected click of the gear locking into place. Instead, they heard a thud. A light didn't illuminate to indicate the landing gear was locked down. What was happening?

The pilot was an experienced veteran with more than 27,000 flight hours. The first officer and flight engineer were experienced as well. Unable to visualize the landing gear underneath the DC-8, the crew contacted Portland tower and received instructions to circle at a low altitude, offloading jet fuel, in preparation for a possible emergency landing. As the aircraft circled for an hour, low fuel alarms began to sound. Flight attendants were preparing passengers for an emergency landing. The captain radioed the tower his plans to land in another 15 minutes. Upon hearing this, the flight engineer responded to the captain with how many pounds of fuel remained and said there was not enough for 15 more minutes. They began to lose engines. At 6:15 pm, they crashed in a wooded area about six miles from the airport. One crewmember and eight passengers were killed, and 23 crew and passengers were seriously injured.

The report of the National Transportation Safety Board (NTSB)[15] determined, "That the probable cause of the accident was the failure of the captain to monitor properly the aircraft's fuel state and to properly respond to the low fuel state and the crewmember's advisories regarding the fuel state. This resulted in fuel exhaustion to all engines. His inattention resulted from preoccupation with a landing gear malfunction and preparations for a possible landing emergency. Contributing to the accident was the failure of the other two flight crewmembers either to fully comprehend the criticality of the fuel state or to successfully communicate their concerns to the captain."

Syed describes this as a watershed event in aviation safety.[16] This may be appreciated understanding two aspects of the NTSB investigation following the Flight 173 crash.

First, when the NTSB issued its accident report in June 1979, it had conducted a thorough and rigorous analysis of the cause of the crash and contributing factors. Syed reports that when the lead investigator interviewed the pilot while he was still in the hospital, the pilot recalled the fuel being depleted 'incredibly quickly.' He was suggesting there might have been leaks in the fuel tanks.[17] Investigators tested the hypothesis with data from United Airlines and Douglas Aircraft Company, as well as their own calculations. These data included the fuel of the aircraft at the gate in Denver, calculations of fuel burn rate, meteorological information, the flight plan, and aircraft monitoring system data. The conclusion noted by Syed is that "the leak was not in the tank, but in (the pilot's) sense of time."[18]

In addition to testing the leaky fuel tank hypothesis, the NTSB investigators carefully documented the chronology of events. This was made possible by recovery of the cockpit voice recorder (CVR) and the flight data recorder (FDR). These technological innovations, commonly known as the "black box," allowed investigators to test hypotheses of what caused the crash using evidence. The investigation also included interviews, review of personnel records, review of the aircraft maintenance and flight history, and a physical analysis of the landing gear assembly, all conducted by a team that had traveled to the crash site to examine the physical evidence. The aircraft had not run out of fuel due to a leak in the tank. The crash resulted from the tunnel vision of the pilot, being task-focused on the landing gear, and the failure of communication within the crew to be situationally aware. As Syed states, "Only through an investigation, from an independent perspective, did this truth come to light."[19]

The second aspect of the Flight 173 investigation that led to a breakthrough in aviation safety was the recognition of an emerging trend. Prior to Flight 173, there were preceding crashes, including Eastern Airlines Flight 401 (December 29, 1972), resulting in the loss of 101 lives, and the collision of KLM-Royal Dutch Airlines Flight 4805 with Pam American Airlines Flight 1736 (March 27, 1977) killing 503. A NASA-sponsored workshop was held in 1979 to examine failures in "interpersonal communication, decision making, and leadership."[20] This helped pave the way for the development of Crew Resource Management (CRM). The impact of CRM moved the culture of commercial aviation cockpits from a captain-as-king mentality to shared accountability for safety using a set of skills and shared safety language.[21]

The systemic improvements in interpersonal communications and decision-making, in combination with continuing technological improvements, is what resulted in an eight-year record of no fatalities. The systemic improvements were made possible by the thorough and credible investigations into each crash, in other words, with RCA.

1.4 LEADERSHIP SUPPORT FOR A CULTURE OF SAFETY

Organizations may have reasons beyond those cited for supporting RCA. Sometimes the motivation is more external than internal. That is, rather than viewing RCA as a tool for learning and preventing harm, administrators with concerns related to licensing and potential sanctions view an RCA as a way of demonstrating the organization's level of response. David recalls a conversation in a health system leadership huddle following the report of a patient fall with significant harm: members spoke of anticipating the arrival of an investigator and urged that a rapid RCA be conducted. Is regulatory risk the reason for doing RCA? Healthcare organizations must determine their vision of a culture of safety, whether for regulatory reasons or as a learning organization committed to increasing reliability and decreasing harm.

According to the report of the expert panel convened by the National Patient Safety Foundation,[22] "The most important recommendation of this report: that leadership (boards/governing bodies as well as executives) must establish a safety culture as the foundation to achieving total systems safety." In turn, RCA is most effective when conducted in an established culture of safety.

1.4.1 LEADERSHIP RESPONSIBILITY FOR RCA

One reason for committing resources to RCA is to enable the sharing of lessons learned and improvements made with caregivers, medical staff leaders, senior executives, and boards. Reports of the number and rate of serious safety events, findings of RCAs conducted, and successes or failures in implementing corrective action plans support leaders in fulfilling their responsibilities to patients and communities.

In healthcare, a strong safety culture is one in which healthcare professionals and leaders are held accountable for unprofessional conduct yet not punished for human errors; errors are identified and mitigated before they harm patients; and strong feedback loops enable frontline staff to learn from previous errors and alter care processes to prevent reoccurrences.[23]

Thorough and credible RCA is an important tool for the learning organization, and part of the overall safety culture.

1.5 PATIENTS AT THE CENTER

Finally, consider those who entrust us with their care. The most important reason for investing in thorough and credible RCA is the expectation of patients that they will not be harmed. Medical errors can have long-lasting and significant impact. Patients and their loved ones expect to receive safe, compassionate care, and be free from harm. When errors occur, they expect us to learn and prevent future harm.

Many healthcare organizations are adopting a target of zero harm. Given the level of complexity of our healthcare systems, RCA alone will not result in zero harm. Systems approaches, including human factors engineering and high-reliability design, are also needed. Evidence of a culture of safety as paramount includes an obsession with possible modes of failure, often described as attributes of a High Reliability Organization. Yet, without the rigor of analysis provided through RCA, we are likely to experience reoccurring unintended events.

Syed tells the story of "A Routine Operation."[24] Elaine Bromiley was a 37-year-old married mother scheduled for sinus surgery. After anesthesia had been induced, the anesthesiologist struggled to place an airway. As Elaine Bromiley's oxygen saturation dropped, the highly experienced clinical team tried different laryngeal masks without success. The anesthesiologist and the ENT physician were joined by another anesthesiologist who realized something was wrong and came to assist. Together, they worked for 20 minutes while Elaine Bromiley's oxygen saturation levels reached 40%. Nurses in the operating suite were concerned. One brought in a tray for performing a tracheostomy. However, the attempts by the nurses to interrupt and voice the concerns were unheard. The physicians had become task-fixated, like the pilot of Flight 173. By the time a decision was made to wake Elaine Bromiley up and her oxygenation was restored, she had suffered a prolonged period of anoxia from which she did not recover.

One of the most remarkable aspects of this story is that Martin Bromiley, Elaine's husband, was an experienced commercial airline pilot. Accustomed to the practice in aviation of rigorously analyzing all accidents, Martin Bromiley pressed to learn the results from the hospital's review. He wondered what lessons were learned and should be shared. The hospital, apparently still caught in an era lacking transparency, was initially uncooperative. Undeterred, Martin Bromiley continued to seek to understand what happened that caused the death of his healthy wife. Ultimately, the hospital agreed to commission an independent review.[1] Even more remarkably, Martin Bromiley wanted to make the report and the Coroner's inquest verdict publicly available. He gave permission to use the report and verdict for the purpose of learning. The title page includes Martin Bromiley's intent for sharing,

"So that others may learn, and even more may live."[1]

In summary, healthcare organizations have good reasons for dedicating resources to RCA, including uncovering the latent causes of errors, caring for caregivers impacted by unintended events by including them in designing improvements, and supporting organizational learning as a key to a culture of safety. RCA is one tool for analysis in a safety management system. The next chapter will examine additional forms of analysis and the need to select the best approach to a problem.

1.6 QUESTIONS TO CONSIDER

1. How do your board directives and organizational goals speak to safety?
2. How are your senior leaders engaged in supporting a culture of safety and the use of RCA?
3. What motivates you to be a leader of RCA?

NOTES

1. D. Berwick, *Escape Fire* (New York: Commonwealth Fund, 2002). Edited version of the Plenary Address delivered at the Institute for Healthcare Improvement 11th Annual National Forum on Quality Improvement in Healthcare, December 9, 1999.
2. N. MacLean, *Young Men and Fire* (Chicago, IL: University of Chicago Press, 1972).
3. S. Ulstrom et al. "Suffering In Silence: A Qualitative Study of Second Victims of Adverse Events." *BMJ Quality & Safety* (2013): 1–7.
4. Sue Scott, University of Missouri Health System. For example, www.hqinstitute.org/sites/main/files/file-attachments/hqi.2016.scott-edit.pdf?1488269098.
5. Medically Induced Trauma Support Services at mitss.org.
6. Resilience in Stressful Events at johnshopkinssolutions.com.
7. Matthew Syed, *Black Box Thinking* (London: John Murray, 2015), 292.
8. NPSF National Patient Safety Foundation, *RCA2: Improving Root Cause Analyses and Actions to Prevent Harm* (Boston, MA: National Patient Safety Foundation, 2015), 11.
9. J. Reason, *Human Error* (New York: Cambridge University Press, 1990).
10. U.S. Institute of Medicine, *To Err Is Human: Building a Safer Health System* (Washington, DC: National Academies Press, 1999).
11. National Patient Safety Foundation, *Free from Harm: Accelerating Patient Safety Improvement Fifteen Years After to Err Is Human* (Boston, MA: National Patient Safety Foundation, 2015).
12. Ibid., 8.
13. Ntsb.gov/investigations/data/Pages/aviation_stats.aspx.
14. Accounts are included in S. Gordon, *Beyond the Checklist* (Ithaca and London: ILR Press, 2013); Syed, *Black Box Thinking*.
15. NTSB Aircraft Accident Report. NTSB-AAR-79–7. June 7, 1979.
16. Syed, *Black Box Thinking*, 26.

17. Ibid., 31.
18. Ibid.
19. Ibid.
20. Gordon, *Beyond the Checklist,* 29.
21. For an explanation of Crew Resource Management see Gordon, *Beyond the Checklist.*
22. NPSF, *Free from Harm,* 11.
23. Syed, *Black Box Thinking,* Chapter 1.
24. M. Harmer, Independent Review on the Care Given to Mrs. Elaine Bromiley on March 29, 2005.

2 Forms of Analysis

Having addressed the question of *"Why?"* to devote the time and resources to analyzing unintended events, we turn to the question of *"How?"* RCA is one tool for analysis, problem-solving, and improvement work. There are many others. And, unfortunately, RCA is not well defined in healthcare. The term can be used without specific regard for the precise definition of root cause analysis, as opposed to other forms of analysis. It is, therefore, incumbent on those vested with facilitating RCAs to help assure it is appropriately applied.

While this is a book focused on and advocating for RCA, it is important to note that there is a risk of over-using or misapplying the tool. Let us look at some of the options and consider the varying contexts in which analysis is needed. We will then address the question of how to decide which form of analysis to use.

2.1 TROUBLESHOOTING

Troubleshooting may be needed and useful at the time an unintended event occurs. As an analytical tool, troubleshooting is utilized in real time or close to real time. In the case of one event, an error in the selection of medication used to perform a nerve block in surgery resulted in extravasation injury of the tissue. The impact of the error was discovered after the procedure was completed and the patient was in the post-anesthesia unit. The post-anesthesia nurse caring for the patient observed the deteriorating tissue and notified the patient's anesthesiologist. Quickly the anesthesiologist, a partner, and the surgeon, came together to brainstorm what might have caused the problem. In troubleshooting this case, they were searching not only for the cause but also how to rapidly reverse the situation and prevent further harm.

This type of troubleshooting is like what is seen in manufacturing when a machine goes down. In the manufacturing setting, the operator brainstorms possible causes with a focus on returning the machine to use as quickly as possible. This type of troubleshooting is done by individuals or teams, most frequently near the time and location an unintended event occurs.

2.2 SIMPLE PROBLEM-SOLVING

Another approach to analysis and problem-solving is the "just-do-it" category. This approach is generally used for simple problems with recognizable cause and effect. Simple problem-solving may benefit from the application of rules such as no financial expenditure, minimal data needed, or three steps or less for resolution. This "just-do-it" approach may be identified and implemented by individuals or teams, and usually occurs sometime after the problem has been identified.

Consider the case of a suggestion raised in a unit or department daily management huddle. If the suggestion can be implemented easily and will have positive impact, the green light may be given to proceed. The assignment can be given to one or more caregivers with a request to report back at huddle in one to three days. The expectation is the solution will be implemented and communicated for team awareness.

2.3 COMPLEX PROBLEM-SOLVING

Other types of failures, like frequent, persistent events or issues that span multiple departments, require more analysis and in-depth problem-solving. Improvement methodologies like Lean or Six Sigma, which incorporate specific techniques for defect elimination, are useful for this type of complex problem-solving. It is important to incorporate the Lean principles of process observations in the workplace (gemba[1]) and the engagement of those who do the work to understand the problem and design the improvements.

This problem-solving approach requires greater time and resources and may take multiple hours or days of process observations and data collection to describe the nature of the problem, followed by hours or days of team-based analysis and design. Solutions are developed by a team of people who do the work from all the areas involved. The benefit of this additional investment is a deeper understanding of the issue, as well as greater team engagement and faster and broader adoption of the solution.

2.4 A3 THINKING

Many organizations using Lean methodology will utilize A3 documents. An A3 is a single 11 × 17 page that describes the problem, the problem scope and background, the goal, gathered data, and a root cause analysis. The A3 also includes a series of planned improvement activities, and the metrics to be measured. It facilitates the process of analysis to understand the roots of the problem more deeply and helps to visualizes the problem-solving process leading to a thorough understanding of the problem before implementing countermeasures.

2.5 5 WHYS

Whether for an A3 or simply doing process observations, many will use the practice of asking "*Why?*" five times as a form of analysis. The 5 Whys move beyond troubleshooting to searching out root causes. John Shook[2] observed for manufacturing that workers may be skilled at getting machines back up and running when a problem develops, but not so good at preventing the problem from reoccurring. Effective application of the 5 Whys requires a commitment of time for inquiry and at least two people (the observer and the person who does the work).

Toyota is credited with developing the technique to identify root causes of problems and to prevent future failures. Taichi Ohno, father of the Toyota Production System, described asking "*Why?*" five times as the basis of Toyota's scientific approach to problem-solving.[3] The reason for asking "*Why?*" five times is based on the recognition that the initial idea about the cause of a failure may not go deep enough.

Individuals or teams may practice asking *"Why?"* five times to better identify the root cause of a problem to be remedied. Ohno provides the following example.[4]

> *Problem:* Machine broke down during operation.
>   *Why?* The shaft broke.
>   *Why?* The machine was running too fast.
>   *Why?* The RPM gauge has been broken a month.
>   *Why?* Maintenance didn't fix it.
>   *Why?* It was scheduled to be fixed next week.
> *Conclusion:* We had a communication problem that, left unresolved, resulted in serious equipment failure.

Consider this healthcare example of the use of 5 Whys. A radiology technician entered the results of an examination on the wrong patient chart. A leader went to learn more and used the 5 Whys technique.

> *Problem:* Two patients arrived from a trauma and were being cared for simultaneously by the trauma team and X-ray technician. The technician X-rayed one patient but entered the results on the other patient's chart.
>   *Why?* The technician opened the wrong patient's chart.
>   *Why?* The technician didn't see any patient wristband identification.
>   *Why?* The ID band was on the ankle and covered by a blanket.
>   Why? The patient registration clerk placed the ID band on the ankle to avoid interfering with the clinical caregivers in the room.
>   *Why?* The unusual number of patients and caregivers in the room restricted access to the patients' wrists.

2.6 FAILURE MODES AND EFFECTS ANALYSIS

Robert Latino makes the case for using RCA in the context of Failure Modes and Effects Analysis (FMEA). FMEA is the disciplined analysis of the potential failures in a multi-step process. It is important to define the process being analyzed, at least at a high level. FMEA involves a team reviewing and confirming the high-level process steps, and then for each step, the team identifies the possible ways in which that step could fail, the frequency of the potential failures, and the negative impact when failures occur. Sometimes how easily a failure is detectable is also considered. The team calculates a risk prioritization number (ranking) associated with each possible failure by multiplying the frequency, severity, and detectability. The critical few failure modes are determined with a Pareto chart. After determining at which steps failures occur and what can trigger the failures, the team turns to RCA to understand the latent root causes of the failure modes more deeply. Action plans are then designed to prevent the modes of failure at the process step in which they occur.

FMEA can be used prospectively for newly designed processes. In this situation, the team cannot rely on failure data. They draw on their experience to envision how the proposed process could fail when operational. The goal is to identify and eliminate as many of the potential failures as possible before implementing the new process. For existing processes, FMEA benefits from the existence of failure data. This is sometimes called an Opportunity Analysis (OA). Opportunity analysis results in a Pareto chart of known failure modes. This allows prioritization of action plans to reduce or eliminate the modes of failure at the process steps where they occur.

2.7 CASE QUALITY REVIEW

In some cases, healthcare leaders may ask for an RCA without being able to identify a specific unintended event. The events of a single case may cause concern with the overall quality of patient care, or leaders may be responding to the dissatisfaction of a patient or family, or caregivers. In these circumstances, it is difficult to perform RCA without an understanding of the event to analyze.

A more appropriate type of analysis for this circumstance may be to perform a case review or a quality-performance improvement analysis. These types of analyses are often structured conversations by multidisciplinary treatment teams. By pulling a team together to arrive at a common understanding of the facts of a case and the progression of patient care, there can be value in this type of retrospective review. It often allows caregivers to reflect on the experience and identify what they would do differently in the future. These reflections may generate important improvements in clinical care.

2.8 APPARENT-CAUSE ANALYSIS

In another healthcare example, a leader noticed a pattern of events involving oxygen (O_2) tanks. Reports of tanks being empty and stored as if they were ready for use had spiked. The leader asked if there an identifiable pattern to the events. This type of analysis is referred to as "apparent cause" analysis. Looking for commonality among the reports can be done by an individual or team in a relatively short time frame, usually within days. This may be important in setting in place a rapid countermeasure. In our example, process observations revealed variation in the storage of O_2 tanks and a lack of clearly defined standardized work processes. Signage was rapidly created to provide visual cues for storage. An improvement team began defining standardized work for the transport and storage of tanks.

2.9 LOGIC TREE

A Logic Tree is a graphic representation of how latent system vulnerabilities led to decisions which, in turn, led to failure. It shows a series of cause-and-effect relationships, supported by evidence. Several software vendors support various formats of fault trees or logic trees. The main benefit of the Logic Tree[5] is the methodology of deductive reasoning.

The logic tree methodology is the principal form of analysis we advocate for RCA teams. However, the method is a resource-intense approach involving the gathering and analyzing of evidence to determine cause-effect branches and to rule hypotheses in or out for how an unintended event occurred. Chapters 3 and 4 will go into more depth on logic tree methodology.

2.10 META-ANALYSIS

The authors were asked by a hospital administrator for a review of all the latent root causes from RCAs since opening a new facility to better understand the impact of the move. A "meta-analysis" was done, looking at the RCAs conducted during a specified period. Each of the latent roots was categorized into affinity groups, such as the lack of having a standard process, inadequate policies and procedures, or communication failures among the team. The categories were compiled and organized into a Pareto chart. Based on the frequency of latent root causes, which indicated the most prevalent system vulnerabilities, we recommended strategic direction for performance improvement. Chapter 9 looks at trending RCA results.

2.11 PEER REVIEW

Many state legislatures have given statutory authority for credentialed clinicians to review one another's cases without fear of legal discovery. Generally known as peer review, the process resides with a medical staff structure or similar "peer review body." Peer review is generally associated with physicians, though some institutions have also created peer review bodies for nurses or other like-credentialed caregivers.

The value of this forum cannot be overstated. Creating an opportunity for similarly credentialed clinicians to review, critique, and learn from cases with errors or poor-quality outcomes supports shared learning and actions that improve quality and safety.

However, the focus of peer review is distinct from that of the RCA. Peer review is intended to review and critique the decision-making of a clinician relative to similarly credentialed clinicians in similar circumstances. It is the forum in which medical/clinical judgment is analyzed.

However, RCA and peer review are not mutually exclusive. In fact, it is not uncommon for both to occur following an unintended event with serious harm. The question is how to optimize both processes without compromising the protection of confidentiality. In our experience, the sequence matters for two reasons. First, the findings of an RCA are generally able to be shared with a peer review body. Peer review results are typically not shared with the RCA team. Further, the analysis of systems and process gaps that led to decisions may provide important insights for the peer review committee. For these reasons, we recommend conducting an RCA in advance of a peer review.

In the case of a heart surgery during which the patient suffered an anoxic brain injury after a medication error, the RCA revealed changes in size and colors of vials of emergency medications used during codes. Those changes, along with the human factors experienced during the high-stress situation of a code, were key to understanding how the anesthesiologist erred in calculating the dose of medication. The peer-review process benefits from including the findings of the RCA.

2.12 MORBIDITY AND MORTALITY CONFERENCE

Physician training involves the practice of reviewing errors and poor-quality outcomes in morbidity and mortality (M&M) conferences. Also protected from legal discovery, these conferences are intended for clinicians to explore causes of morbidity and mortality. The focus is on the pathophysiology as well as the care. This again differs from the focus of RCA, though it may be complimentary. Understanding the pathophysiology may be important to understanding the sequence of events and can inform RCA chronology.

In the case of a patient with chronic obstructive pulmonary disorder who became over-oxygenated on respiratory support, it was important to understand the physiology of O_2 and SPO_2 on the respiratory drive. Understanding this allowed correlation between the oxygen therapies and decisions regarding treatments, the focus of the RCA.

2.13 SUMMARY OF THE DIFFERENT FORMS

As evident, there are many tools available for analyzing failures. We can organize these tools along the dimensions of how much time they require, the strength of the gathered evidence, and the effectiveness or accuracy of their results. There is a hierarchy of these tools based on the time and resources required. These are important considerations in healthcare.

Healthcare administrative leaders may lack a working knowledge of the variety of analytic tools. We have observed that many leaders may know the term "RCA" and use it generically. The leader is motivated to have an analysis conducted, to capture lessons learned, and to prevent reoccurrence of an unintended event. They express the *why* to investigate. It is often the responsibility of the skilled RCA facilitator to advise *how* best to accomplish the objective.

2.14 SELECTING THE FORM OF ANALYSIS

Selecting the appropriate form of analysis and when to use RCA are important considerations. A thorough and credible RCA and action plan require intensive use of resources, specifically leaders' and caregivers' time. Knowing when to use root cause analysis and when to use other forms of analysis provides good resource stewardship.

A prioritization tool can help teams navigate these decisions. Table 2.1 shows a simple matrix based on the anticipated frequency of an unintended event and the most likely harm that would result should the event occur. The tool helps stratify and prioritize events for analysis, which leads to recommended approaches for analysis and improvement.

Table 2.2 suggests tools for analysis based on the prioritization level. "Red" events are generally those requiring rigorous analysis and mitigation plans. These are the serious harm events, as well as events that should never happen in healthcare regardless of harm level[6], and the high-risk close call events. "Orange," "yellow," and "white" events may benefit from RCA but may also be adequately addressed by other less intensive forms of analysis.

One pitfall for teams prioritizing can be inaccurately defining the unintended event to be prioritized. Take, for example, a medication error involving an infrequently

TABLE 2.1
Prioritization Tool

4 Catastrophic Harm	Permanent harm Loss of body function or disability requiring life-sustaining treatment	ORANGE	ORANGE	RED	RED
3 Serious Harm	Significant intervention Injury lasting < 6 months	YELLOW	ORANGE	ORANGE	RED
2 Minor Harm	Minor treatment No significant intervention	WHITE	YELLOW	ORANGE	ORANGE
1 Negligible Harm	Assessment or monitoring to preclude harm No or minimal intervention	WHITE	WHITE	YELLOW	ORANGE
Event Severity		Less than 1–2 times/year	Close to Monthly	Close to Weekly	Close to Daily
	Event Reoccurrence	1 Unlikely Occurrence	2 Possible Occurrence	3 Likely Occurrence	4 Almost Certain Occurrence

administered chemotherapy medication. Given the level of risk, this might be prioritized as a red event. However, to establish a priority level, a team needs to agree what "event" is being classified. Imagine the difference if the team considers medication administration as the event (high frequency of occurrence and low probable harm level) versus chemotherapy administration as the event. The answer will impact the weighting of the prioritization. It is important to be specific when defining the unintended event to be prioritized.

Other pitfalls involve how the questions of frequency and expected severity of harm are framed. Frequency should be thought of as the number of times the process is performed. In other words, how often is there a chance of failure? In some cases, teams may also want to consider historic data on how often failures occur. Another risk is how we evaluate the potential harm of an event. In healthcare it is often conceivable to think that in many potential unintended events the possible consequence of failure is death. This will not help stratify the potential analyses and improvement work. It is better to ask, "What is the most likely clinical outcome?" The value of defining the event to be prioritized is in the impact on the decision of how much resource to devote to an analysis, that is, in selecting the right form of analysis.

In summary, RCA is one analytical tool among many. Applying the right level of resources for analysis and improvement work may depend on the RCA facilitator

TABLE 2.2
Hierarchy of Analytical Tools

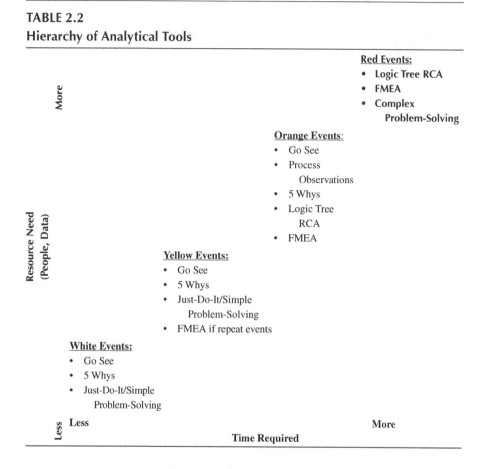

helping to guide prioritization and selecting the right tool to use. There are important reasons for devoting the intensive resources to RCA; healthcare organizations do well to define the circumstances in which RCA is required, those in which it is advised, and those circumstances and events better suited to other types of analysis and improvement work.

2.15 FORMS OF RCA

A strong knowledge of the variety of analytical tools available and how to prioritize which cases need RCA does not answer the question of which of those methods to use for a thorough and credible RCA. The following case study shares the journey of the authors in adopting the PROACT Logic Tree methodology[7] as our standard for RCA.

2.15.1 TELEPHONE CUTOVER DISASTER: A CASE STUDY

A Level Two trauma center scheduled a conversion to a new telephone system on a Friday afternoon. Unfortunately, the cut-over to the new system failed immediately, and even more unfortunately, once the new system was activated it was not possible to flip back to the old system. Neither the vendor nor the hospital facility technicians were able to successfully troubleshoot a quick recovery and the medical center was forced to operate on a cobbled together emergency telephone system for an extended period.

Harold was asked as a Quality Department Improvement Engineer to conduct an analysis of the failure. Harold started with a series of probing interviews with key stakeholders and organized the findings into a Fishbone (Ishakawa) diagram with categories of Planning, Procedures, Policies, People, Environment, and Equipment. His results were extensive, with 63 causes identified and 3–4 contributing "Why" statements for each cause.

At about this same time, David transitioned from being the manager of a busy inpatient psychiatric unit into Risk Management. As a clinical manager, David had been in RCAs but was dissatisfied with the traditional healthcare RCA process and results. David observed it was not uncommon to approach the end of a two-hour review of an unintended event, having discussed a set of factors considered important by an accrediting body, to hear a caregiver say, "Can we just talk about what happened?"

A presentation at ASHRM[8] that year described the difference between a shallow cause analysis and root cause analysis. Robert Latino presented a methodology called the Logic Tree. Latino brought decades of manufacturing experience with RCA and reliability engineering to the healthcare arena. His presentation contrasted forms of RCA using a healthcare case study. This presentation inspired Harold and David to apply the Logic Tree, along with several other methods, to the telephone failure. Using the same set of facts, we were able to compare and contrast several methodologies, as described in the following section.

2.15.2 5 WHYS, FISHBONE, AND BIMODAL METHODS

Those familiar with the Toyota Production System and Lean often practice 5 Whys. The 5 Whys method of RCA involves going to the workplace and inquiring about the roots of a failure. As described earlier in this chapter, the practice is to interview and continue asking "*Why?*" an unintended event occurred until arriving at the deeper root cause. 5 Whys is usually done by an investigator interviewing one operator. This brings inherent limitations. For one, the inquiry can only go as deep as the knowledge of the operator. It also reflects the insights of the one operator. These insights may be supported by evidence, but often with this method the investigator accepts the representations by the operator without asking for underlying evidence. 5 Whys, therefore, is prone to opinion and bias. The major deficiency of 5 Whys, however, is its inherent inability to determine multiple roots. By definition, it drives to only a singular root. 5 Whys was never designed, or intended to be used, as an RCA methodology for complex and broad analysis. The 5 Whys method was intended to be a conversational technique in the workplace to teach people to dig deeper to gain an understanding.

The Ishikawa or Fishbone diagram uses categories and brainstorming for causes organized by category that may have contributed to the unintended event. The categories are not related to the actual cause-and-effects of the event, but rather they try to help assure people "not miss something" as they brainstorm potential causes. Traditional categories are the 4Ps (people, process, policy, procedure) or 4Ms (methods, materials, manpower, machines).

But the Fishbone diagram does not describe the hierarchy of causes (cause-effect logic chain) needed to identify the systemic latent causes from which human decisions and consequences arise. On a Fishbone diagram, "causes" are a mix of physical, human, and systemic contributors, with no visualization or linkage of cause and effect from latent root to event. In the case of the Telephone Cutover RCA, with over 200 "Whys" of apparent equal importance, it was not possible to identify where to start in developing action plans.

One remedy for this limitation of the Ishikawa or Fishbone diagram is to combine it with the 5 Whys. Using this approach, "causes" listed in each category have "*Why?*" asked 5 times to drill to a lower root. However, while this approach is better at identifying systemic root causes, it still does not ensure all the causes are identified. Some practitioners have expanded the Fishbone categories to have 8Ps or some combinations of the Ps and Ms plus other categories (as used in the Telephone Cutover Fishbone) in an attempt to be comprehensive. The 5 Whys are also difficult to draw on a Fishbone diagram as one runs out of space very quickly, leading to tables or other workarounds being used, thereby losing the benefit of the Fishbone visualization. The limitations of the 5 Whys method previously described also remain.

In this same period, the Healthcare Advisory Board was teaching a bimodal RCA methodology that expanded the 5 Whys to allow two branches from each "*Why?*" which resulted in 16 latent roots versus the 5 Whys singular root. This method was tested with the Telephone Cutover evidence and correctly identified a small sub-set of the latent roots (16 of the 76), better than the 5 Whys but still clearly inadequate.

In comparison to the 5 Whys, Fishbone, and Bimodal 5 Whys, the logic tree technique allowed definitive cause-effect placement. This led to the clear determination of 76 systemic latent roots, assignable to vendor, customer, and joint. In addition, clearly revealed was where and how the same systemic latent root caused multiple unique contributing effects. See Figure 2.1 to understand the breadth of the Telephone Cutover Logic Tree.

Through this comparative analysis, we determined the logic tree to be the RCA technique most capable of defining all the latent roots and identifying the cause-and-effect relationships of those latent roots leading to the unintended event. The result was a decision to further test the expanded Logic Tree methodology as presented by Reliability Center Inc. with clinical cases.

In one case, the Logic Tree identified physical environment factors that enabled a patient to elope from a psychiatric unit, leading to a redesign of the unit entry. Other methods may have looked at policies and procedures and resulted in reminders and training for caregivers, missing the structural barriers to safety. In another case, an RCA of a wrong-level spine surgery discovered the vulnerability created by the ways surgeons and radiologists independently count vertebrae. Other methods may have 'blamed' the spine surgeon for an unpreventable mistake. After a series of RCAs on serious safety events using the Logic Tree, a decision was made to standardize the RCA process using the Logic Tree methodology.

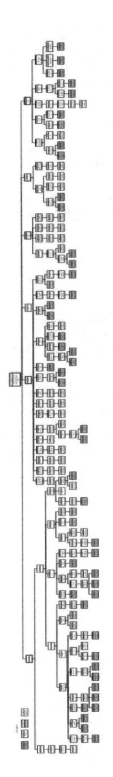

FIGURE 2.1 Logic Tree of Telephone Cutover

Source: PROACT image used by permission of Reliability Center, Inc.

In summary, there are numerous RCA methods and software available. The intent here is not to suggest there is only one method of RCA to use in all cases. Rather, it is to share the journey and decisions made that led to our adoption of a standard RCA approach for complex and multi-discipline unintended events.

2.16 QUESTIONS TO CONSIDER

1. What analytical and improvement tools does your healthcare organization use?
2. Are there additional tools you might want to add to your toolbox?
3. How are events prioritized for review?
4. Would you describe your RCA techniques as rigorous and robust?
5. Are you confident you're getting to systemic latent roots in your RCAs? And how would you know?

NOTES

1. "Gemba" is the Japanese term used in the Toyota Production system referring to *the place where value is created.*
2. J. Shook, "Bringing the Toyota Production System to the United States," in *Becoming Lean,* ed. B. Liker (Portland: Productivity Inc., 1998), 58.
3. T. Ohno, *Toyota Production System: Beyond Large-Scale Production* (Boca Raton, FL: CRC Press, Taylor & Francis, 1988).
4. Ibid., 17.
5. The authors have utilized the PROACT Logic Tree and are indebted to Robert Latino and Reliability Inc. for teaching and advocating for rigorous RCA in healthcare.
6. National Quality Forum (NQF), *Serious Reportable Events in Healthcare—2011 Update: A Consensus Report* (Washington, DC: NQF, 2011).
7. R. Latino, *Patient Safety: The PROACT® Root Cause Analysis Approach* (Boca Raton, FL: CRC Press, Taylor & Francis, 2009).
8. Robert Latino, *Shallow Cause Analysis v Root Cause Analysis* (San Antonio, TX: ASHRM, 2005).

3 Pre-Work for an RCA Team Meeting

To ensure a productive RCA team, the facilitator has several actions to take leading up to the RCA team meeting. These should be done in coordination with the Executive Sponsor for the RCA, and the Process Owner (more about these two roles in Chapter 7). This chapter looks at some of the immediate needs following an unintended event, tasks in preparation for the team meeting, and the logistical needs for the RCA team meeting.

The decision to move forward with an RCA initiates a series of actions. Latino[1] uses the acronym "PROACT" to describe the sequence of Preserving information, Ordering a team, Analyzing root causes, Communicating action plans, and Tracking results. This acronym roughly outlines the sequence of the major tasks.

3.1 EARLY INVESTIGATION AND PRESERVATION OF EVIDENCE

First, we must consider the immediate response to an unintended event that has just occurred or been discovered. The term "RCA" is sometimes used synonymously with a team meeting for conducting an RCA. But it's a mistake to fully equate the two when, in fact, a successful RCA must begin when an unintended event is discovered. In either case, it should trigger a series of steps ultimately leading to uncovering the causes of the event and what can be done to prevent its reoccurrence. In other words, RCA is not just a meeting. *RCA is a rigorous, standardized process designed to lead to tangible results for preventing failures and improving safety.*

Thinking this way, we can begin to identify the high-level steps in the RCA process. The steps begin with what needs to happen immediately: What is the path when an unintended event is discovered? Who is informed and how? In acute care settings, unintended events are often discovered immediately or within a short time: A medication error results in observable changes to the patient's condition, or the failure of an oxygen delivery system is discovered through the desaturation of a patient. In many cases, though, there is a gap in time between the error and the discovery. For example, a failure to correctly reconcile medications on admission to the hospital that results in missed orders for a critical anti-seizure medication may not be evident for many days, until a patient experiences a seizure. An object unintentionally retained after surgery may not be known immediately. In ambulatory settings, it is not unusual for unintended events to be discovered after the patients have left the clinic. An incidental finding on an exam or negative test result may not have been communicated and is not discovered until the patient returns for unrelated care sometime in the future. Discovery of a telephone triage error may not be known for some time. It is important to define a notification process that covers all scenarios, no matter when an unintended event is discovered.

FIGURE 3.1   First Steps in the RCA Process

Immediate notification that an unintended event has occurred or has been discovered allows for the rapid initiation of an investigation, the next major step in the RCA process. (See Figure 3.1 for the initial process steps.) It also allows placement of immediate countermeasures, helping ensure the same event will not happen to the next patient. This may be a temporary bandage until an action plan is developed through an RCA, but the practice is key to ensuring the harm will not happen to another patient while allowing the time needed to proceed with an investigation.

First responders to an unintended event should take charge of sequestering any involved evidence, which can be categorized as parts, people, paper/digital, and paradigms. Take equipment offline and "red tag" it for later analysis if possible. If the equipment is essential and needed to return to service immediately, preserve as much information as possible. This could be done by photographing settings, printing a report, or documenting observations. It can also be important when sequestering equipment to document the position in which it was located at the time of the event. Biomedical engineers may need to conduct an analysis. They can provide the subject matter expertise of the equipment's performance and possible causes of failure. They often have manufacturing information and can be a liaison to a manufacturer's product rep. They can provide data on the equipment's history of use: when it was purchased and brought on-line, its maintenance records, and any prior defects. In some cases, an external laboratory may be needed to analyze a piece of equipment. David investigated a "smoke event," which resulted from a lighting fixture dripping molten material onto a day bed below. The fixture was packaged and sent to an independent forensic laboratory. At stake in the analysis was the safety of all the similar fixtures throughout the hospital. The type of laboratory testing of the wiring and other materials was beyond the capability of the hospital.

Supplies should be treated in a similar fashion. Sequester any involved supplies if possible, being cautious about biohazards. Supply chain leaders can provide valuable information on the products: purchase histories, how widely and frequently the product is used, when the product was introduced, and any product recalls or FDA reports. Your partners in gathering product information for the RCA team might include supply chain leaders, buyers, value analysis leaders, and clinical educators. Bring sequestered supplies from the event, if possible, as well as unused samples of the same product, along with product information sheets, to the RCA team meeting for examination. The show-and-tell can make it easier to understand how the product functions to support the analysis. If the actual products are not available, images may be used. David facilitated an RCA involving a gastric tube incorrectly accessed by a bedside nurse. The supply chain leader found the tube was rarely used. It contained three ports instead of the typical two. The tube had been selected for surgical placement; without explanation it was not familiar to the floor nursing staff. The clinical educator identified the gap in needed training. This coordinated data gathering led the RCA team to limit the types of tubes purchased, and to create laminated one-page just-in-time tip sheets for infrequently used tubes.

Medications should also be considered for sequestration. Medication errors, unfortunately, are often not discovered until they reach the patient. This can present challenges: Is the medication vial or intravenous (IV) bag still safely accessible? Taking involved medication pumps offline and preserving their settings for analysis is important. Equally important is to preserve the vial or bag for future analysis. An assay of the contents may be needed in some cases. Pharmacy leaders are valuable experts and can assist with analyzing the parts preserved from the case, as well as providing information on pharmacokinetics, drug interactions, recalls or black box warnings, and Institute for Safe Medication Practices (ISMP) reports on medication safety.

In addition to preserving data from parts, there are two critical aspects to preserving information from the people involved. One is the interviewing skills required of an RCA facilitator. A second is the emotional state of involved caregivers and potential needs for support. Let us start with tactics for successfully interviewing caregivers.

Edgar Schein[2] advocates leaders adopt an attitude of humble inquiry. The same is needed for RCA facilitators. The objective is to learn what the caregiver knows and understand their perspective. Analysis will come later. Find a private space if possible and introduce yourself: set the stage by sharing the purpose of the interview and your role as an RCA facilitator. There may be questions about confidentiality that you need to be prepared to address based on your healthcare organization's policies. An interview can often begin with an invitation such as, "Tell me about what took place." Open-ended questions elicit more information than do closed "Yes/No" questions. In the interview, you are not only looking for facts, what happened, but the decision-points as well. (See Table 3.1 for RCA Interview Quick Tips.)

TABLE 3.1
RCA Interview Quick Tips

- ☐ Quiet/Private location
- ☐ Introduce yourself
  - ☐ Name
  - ☐ Role
- ☐ Describe confidentiality
- ☐ Ask open-ended questions
  - ☐ Can you tell me about what took place?
  - ☐ What was your understanding?
- ☐ Support the caregiver
- ☐ Answer questions
- ☐ Describe next steps

When asking a caregiver to walk you through the sequence of events, you will come to points when you need to ask, "Why? (Human Root) did you decide to do (or not do) that?" "What led you to that decision?" "What was your understanding?" These types of questions help identify Human Roots. They also give clues to the underlying Latent Roots. Consider questions that elicit information on processes. You might ask, "What usually happens?" "What are the steps you normally follow?" and "How is this situation different?"

The initial caregiver interview provides other opportunities as well. One is to introduce the subject of an RCA team meeting. If the caregiver has not previously attended an RCA team meeting, there may be questions. It is not uncommon for caregivers to have concerns about participating in an RCA team meeting. They may be fearful of being judged by their peers, being blamed, or that their employment is at risk. The interview is an opportunity to describe the purpose of the RCA team meeting and what to expect. Our experience has been that having the ability to interview vulnerable caregivers in advance reinforces the psychological safety of the RCA team meeting. By establishing a humble, supportive relationship with the caregiver, you are providing reassurance that they will be safe.

You may discover that some caregivers are not yet emotionally prepared to participate in an RCA team meeting. Some healthcare errors can cause distress, even trauma, for those involved. It is critical that the emotional well-being of caregivers be assessed and addressed prior to the RCA team meeting. The purpose of the RCA is to provide analysis and action planning. Critical incident defusing or debriefings, spiritual care, peer support, employee assistance, and the support of caregivers' direct leaders are all tools to consider prior to the RCA team meeting. The risk is that emotionally unprepared caregivers, who have not had the chance to address their emotional impact, will be unable to contribute to an analysis. They simply are not yet ready. On the other hand, those who have been able to care for their emotions have almost universally reported that participating in the RCA team meeting and contributing to preventing a repeat of the unintended event is highly valuable. They often express a sense of relief and appreciation for being respectfully included.

3.2 TASKS IN PREPARATION

The recommended steps for the initial investigation may be started by an initial responder and handed off to an RCA facilitator, or be performed by the RCA facilitator directly. Either way, there are subsequent tasks that fall to the RCA facilitator. The next step in the process involves clearly defining the Event, prioritizing the need for further investigation, and determining whether to proceed with a formal RCA, for which a method of prioritization was suggested in Chapter 2. As the RCA team meeting approaches, an event chronology and the Logic Tree Top Box are prepared (see Figure 3.2).

FIGURE 3.2   Steps in the RCA Process

3.2.1 Defining Serious Safety Events

Unintended events are generally classified as a Serious Safety Event, a Close Call or Near Miss, or possibly a Good Catch. Other terms are also used, such as Sentinel Event[3] or Serious Reportable Event[4]. Prioritization will guide which types of events will be analyzed with formal RCA. Do not underestimate the importance of understanding whether and how an event fits the criteria for a Serious Safety Event or Close Call. Leaders and caregivers may have an initial reaction to an event and initially believe there should be an RCA. We have found that it is an important exercise for a multidisciplinary team to review the known facts and decide whether and how the event meets the organization's criteria for conducting RCA.

The National Quality Forum (NQF), the Joint Commission (TJC), and others have determined lists of the types of healthcare errors that should never happen. These include the lists of 29 preventable events[5] by the NQF, and Sentinel Events[6] by the TJC. The American Society for Healthcare Risk Management (ASHRM)[1] has proposed a general definition, "A Serious Safety Event (SSE), in any healthcare setting, is a deviation from generally accepted practice or process that reaches the patient and causes severe harm or death." This definition covers errors of *omission*, as well as *commission*.

The decision of whether to proceed with an RCA must be supported by using accepted criteria and guidelines. Clearly defining the type of unintended event based on the criteria is a prerequisite for successfully constructing a Logic Tree.

3.2.2 Ordering the Team

Information from the initial investigation will guide who is invited to serve on the RCA team. There are several considerations in composing an effective RCA team. Begin with identifying caregivers who have direct knowledge of the events. This may include caregivers absent at the time of discovering an error but who were previously involved in the patient's care. In other cases, this may include subsequent caregivers who responded to the situation and may provide information on the impact of the error. In some cases, the physician is an important RCA team member. In other cases, the physician's participation may not be needed. The RCA Executive Sponsor and the clinical Process Owner can help make this determination.

In addition to caregivers directly involved, other members may be invited related to their role or expertise. See Table 3.2 for a checklist of possible RCA team members.

Leaders in the clinical area of the event should be included. Their presence can show support for their caregivers. (In some cases, you may need to coach frontline leaders or executives on the need to support an atmosphere of psychological safety and humble inquiry.) They are also essential to approving action plans, assuring their implementation, and spreading the lessons learned. Good practice is to include a frontline leader, often the manager, to serve as the Process Owner. The clinical leader can also provide subject matter expertise on the clinical aspects of the case and best practices. An Executive Sponsor is required for every RCA conducted and is most often the Chief Medical Officer, Chief Nursing Officer, Chief Administrator, or Operating Officer. In some cases, this role can be delegated to a Director. Their job

TABLE 3.2
Checklist for RCA Team Membership

(participation based on need)

Core Team

☐ RCA Facilitator
☐ Executive Sponsor
☐ Process Owner
☐ Caregivers
☐ Clinical/Physician leaders

Consider Additional Members

☐ Safety Officer
☐ Risk Manager
☐ Quality Leader
☐ Clinical Educator
☐ Informaticist
☐ Value Analysis Leader
☐ Supply Chain Leader
☐ Pharmacist
☐ Clinical Engineer
☐ Patient Experience Professional

Special Consideration

☐ Patient/Family Advisor
☐ Contracted Services Provider/Rep
☐ Vendors

is to support the focus of the RCA being on improvement and not on blame, remove barriers to successfully performing the analysis, and ensure the successful implementation of the action plan (more on the roles and responsibilities of the Process Owner and Executive Sponsor in Chapter 7).

Your organization may develop a list of caregivers to be included as part of a standardized RCA team. For example, you may consider inviting safety officers, quality leaders, performance improvement specialists, clinical informaticists, risk managers, clinical educators, medical directors, division chiefs, or medical staff leaders. In some cases, participants should include environmental services, supply chain, information technologies, pharmacy, finance, biomedical engineering, and patient experience representatives.

Special consideration is needed with certain potential RCA team members. These include patients/families, patient/family advisors, contracted services providers, and vendors. Risk, legal, and privacy advisors should be involved in deciding whether and how to involve these potential RCA team members. These scenarios are best considered in advance and documented to provide guidance going forward. Decisions are needed on how confidentiality is handled, how protected patient information is shared, and agreements on prohibiting information from the RCA to be used in litigation or responding to a claim. Special circumstances may need to be addressed on a case-by-case basis.

Examples of potential external RCA team members include surgical reps present in the operating room at the time of an unintended event, emergency medical and transport services involved in the care or transport related to an unintended event, or manufacturer reps for equipment or supplies pertinent to the RCA. If there is a shared interest based on the mutual involvement in the patient's care, there may be significant value in reaching an agreement that allows mutual participation in the RCA. At times, the lack of access to the external partners can limit the availability of data and hinder the verification of hypotheses in the Logic Tree. Sometimes this can be countermanded through other sources of data. In the case of a cardiac stent that failed to deploy properly, the manufacturer declined to provide data regarding failure rates or the number of similar cases. However, information was obtained on product defects through a database of publicly reported cases.

Decisions about whether and how to include a patient and/or their family in an RCA are obviously important. Some have advocated for including the harmed patient and family members. In some cases, the patient/family has been invited to attend the beginning of the RCA team meeting. This gives the team the opportunity to hear their recollection of the events and understanding of the consequences. The patient/family may or may not stay to participate in the analysis. Each organization needs to delve into the complex choices involved and make the appropriate decision for each RCA. Patient/family participation in the RCA team meeting may or may not fit on an individual basis. Healthcare organizations need to develop the policies and procedures needed for apology and transparency, whether through early discussion and resolution programs or by other means.

Whether or not the patient/family is invited to attend the RCA, they must be considered as important sources of information for the RCA team to collect and preserve. This may be accomplished by the RCA facilitator interviewing the patient/family. This may need to be done jointly with a clinical leader, executive representative of the organization, a risk manager, or others responsible for follow-up to a harmful unintended event. A similar approach to that taken with involved caregivers is helpful. Introduce yourself, explain the purpose of the interview and the process of the RCA, discuss whether notes may be taken. Address any questions about confidentiality and how the information shared may be used. Invite the patient/family to share their experience. Again, ask open-ended questions. The purpose of the RCA facilitator's interview is to increase understanding of the event. Being aware of the emotional needs of the patient/family is critically important as well. Be careful not to cause further harm with statements or implications. Don't promise anything unless authorized to do so. In many cases, sharing the results of the RCA and action plan is important to the patient/family. They want to know that their suffering has been taken seriously, and steps to prevent others from experiencing the same healthcare failure are underway.

An alternative to inviting the patient/family impacted by the unintended event is to involve a trained group of patient/family advisors (PFAs). Healthcare organizations must take steps to keep the patient at the center, including at the center of

RCAs. Many organizations draw on the experience of PFAs and councils (PFACs). Typically serving as volunteers, these PFAs bring a wealth of experience. Sometimes the experience is borne out of the experience of harm at the hands of healthcare. Their insights into the experience of suffering and pain caused by healthcare failures can help ground an RCA team in the most important reason for convening: patient safety.

We have had some success with carefully selecting PFAs to serve as RCA team members. RCA facilitators, in conjunction with the leader for the PFAC, volunteer coordinator, human resources, or whoever has responsibility for the PFAs, can invite carefully selected individuals to participate. The qualities needed to serve in this kind of role include excellent communication skills, both listening and speaking; the ability to feel comfortable in highly charged emotional settings; the capacity to speak up, including in the presence of physicians and others with significant authority; a commitment to confidentiality; and a strong desire to support safer care. An orientation to the RCA process for PFAs in advance of serving is important. Describing the role and expectations of the PFA helps bring clarity to the PFA and the RCA team. It may be helpful to develop PFA teams to serve on RCAs. Inviting two PFAs to the RCA team meeting allows them to support one another, and the opportunity for them to debrief with each other. The RCA facilitator should debrief with the PFAs, which allows for ongoing learning about the RCA process, as well as monitoring for any concerns or distress the PFAs may be experiencing.

3.2.3 THE CHRONOLOGY

One can consider Logic Tree analysis as viewing an unintended event in reverse, tracing backward in time from the event, frame-by-frame, to the Latent Roots. The forward streaming of the facts is provided by the chronology of the case. The chronology is often prepared by the RCA facilitator, but may also be prepared or contributed to by the clinical leader or content expert. It is important to collaborate on the scope of the chronology. One RCA team failed because the chronology did not go far enough back in the patient's care. This was compounded by the absence of caregivers from the unit that provided care upstream from the place of the unintended event. Deciding where to begin the chronology is crucial; equally important is deciding how much detail and what data to include. It's insufficient to replicate the medical record verbatim. The goal of the chronology is to create a record of the facts that the RCA team can agree on, considering the reality that the involved caregivers may only have knowledge of their own involvement with the patient. Caregivers weave in and out of contact with patients: a chronology that aligns the multidisciplinary involvement helps bring into focus the interactions among the caregivers. The chronology helps get the RCA team on the same page with an understanding of the facts.

The format for chronologies may vary. The goal is to visually convey key facts in a chronological sequence but how the data is displayed may depend on the type of information and the case. The chronology can be built around data extracted from the medical record, from elements from interviews, and from other sources of collected data. Consider the use of spaghetti diagrams, or floor plans, to display the movement of people and parts over time. Some RCA facilitators use Word documents or Excel tables; some prefer Visio flow diagrams.

3.2.4 THE LOGIC TREE TOP BOX

Once data collection is complete and the chronology prepared, it is time to create the Top Box of the Logic Tree. The Logic Tree form of RCA is designed to answer three questions: *why* the event is being reviewed, *what* the basic facts are, and *how* the event occurred. Answers to the first two questions are depicted in the Top Box of the Logic Tree, shown in Figure 3.3.

The "Event," listed at the top, is the undesired result, failure, or problem: the unintended event. The Event may be either a single occurrence or repeated failures. In healthcare, this field indicates whether the unintended event is a Serious Safety Event, Close Call Event, Operational Event, or Quality Improvement Opportunity. It is *why* the RCA is being conducted, and the rationale for the resources dedicated to the investigation and action plan. For purposes of trending, it is recommended to track the types of Top Box events: Serious Safety Event, Close Call, Operational, or Quality Improvement Opportunity. (Trending is discussed in Chapter 9.)

Underneath the Event is the Mode. The early investigation, interviews, data gathering, and preparation of the chronology all feed into the framing of the Mode. A well-defined Mode serves two purposes. First, it indicates which of the organization's categories of unintended events qualify the case for RCA – for example, "Patient Fall" or "Unintentionally Retained Foreign Object." This is the reason for

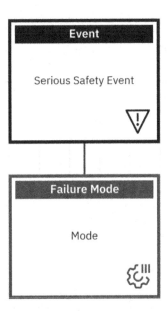

FIGURE 3.3 Logic Tree Top Box

Source: EasyRCA Image used by permission of Reliability Center, Inc.

conducting the RCA and is referred to as the Mode Category: the definition of the type of healthcare failure. Second, the facts, at a high level, are displayed in the Mode using a short statement to provide factual details. The Mode Statement describes what happened in a more specific way than the Mode Category alone. For example, rather than just writing "Patient Fall," which is a Mode Category, a specific Mode Statement could state, "Patient found on the ground between bed and bathroom." Another example, rather than just the Mode Category of, "Unintentionally Retained Foreign Object," a good Mode Statement to guide the team would be, "Retained Lap Sponge after Open Abdominal Surgery." Successful facilitation of an RCA requires accurately framing the specifics of the Mode into category and statement. We have found the Mode Category itself does not provide enough factual detail to focus the RCA team's analysis. Figure 3.4 shows how the specific Mode Statement and the Mode Category are included in the Top Box in PROACT software.

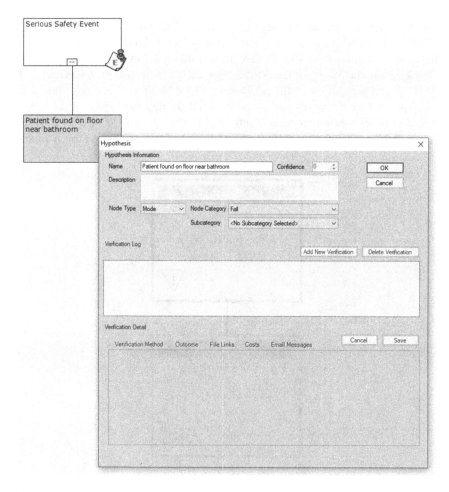

FIGURE 3.4   Mode and Mode Category in Top Box

Source: PROACT image used by permission of Reliability Center, Inc.

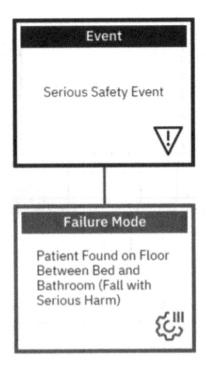

FIGURE 3.5   Top Box with Mode Statement and Mode Category

Source: EasyRCA image used by permission of Reliability Center, Inc.

Another option is to list a Mode Category along with the Mode Statement in the Top Box. See Figure 3.5 as an example in EasyRCA software.

For purposes of trending (see Chapter 9), you want to record the Mode Category. To guide the RCA team, you want to record the Mode Statement.

We recommend constructing the Top Box prior to the RCA team convening. The specific Mode Statement for the Top Box is usually framed by details identified during the early investigation. Clinical leaders and executive sponsors for RCA may confirm the proposed Mode. Generally, the Event and Mode should not be debated in the RCA team meeting, though the team may be asked to confirm them.

3.2.5 TOP BOXES WITH TWO MODES

In some circumstances there may be two Modes in a Logic Tree Top Box. In healthcare, there can be the need to review the response to an unintended event, as well as the cause of the unintended event itself. For example, David led an RCA team in the review of a delayed neonatal resuscitation. The RCA dug into the causes within the Labor and Delivery unit that led to the delay in initiating the code. In addition, during the code response the neonatal resuscitation team experienced gaps in their delivery of the code. Both the delay in initiating emergency procedures, and the delay in carrying them out, impacted the patient. A second Mode was added to the Logic

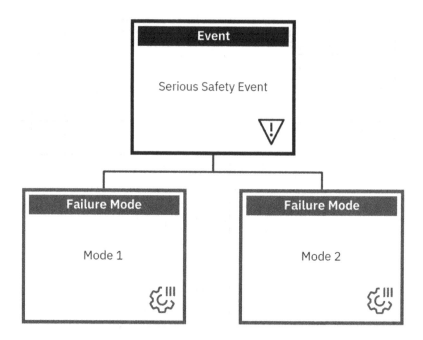

FIGURE 3.6   Logic Tree with Two Modes

Source: EasyRCA image used by permission of Reliability Center, Inc.

Tree to understand the causes of the delays in the response. On the Logic Tree, one Mode referred to the delay in initiating the code, and the second Mode referred to the gaps in the response. Two Modes provided the correct guidance to the RCA team. (Figure 3.6 illustrates two Modes within a Top Box.)

3.2.6 Logistics

The logistical aspects of the RCA team meeting are challenging, and from lessons learned, we developed mitigation strategies for some of the most common logistical issues. Scheduling can be one of the biggest challenges. Coordinating calendars of executives, providers, and frontline caregivers on an ad hoc basis can be a nightmare. It often results in delays of weeks, or even months, before the RCA team meeting can occur. By this time memories have been impacted, the sense of urgency is lost, and it becomes a struggle for the RCA facilitator to motivate the team. As a countermeasure, we found that scheduling RCA block times in advance is the only solution. We shortened the delay to the RCA team meeting by having dedicated days and times on the calendars of busy executives, and with conference rooms reserved. Two to three blocks per week may be needed, depending on the volume of RCAs. When an unintended event occurs and the decision is to conduct an RCA, it is scheduled into the next open block. If there is no RCA, the block can be released.

Scheduling RCA block times requires deciding in advance how much time is to be reserved for an RCA team meeting. We have found it important to block at least three hours. Some cases will not need that time to complete. But it is easier to release the team than to add time. This level of commitment usually requires senior leaders to support the prioritization of time for RCAs and the establishing of the blocks to accomplish the targets for timely completion.

A standardization decision required is whether to gather the team once or for multiple meetings. A lesson learned is the challenge of attempting to hold multiple meetings for completion of an RCA. Only a few cases may be so complex that the analysis cannot be accomplished in three hours. Harold and David jointly facilitated an extraordinarily complex case that required three multi-hour sessions. But this isn't the norm. Our experience is that three hours is sufficient to complete an RCA and develop an action plan.

Ideally, you will have access to conference rooms conducive to team meetings. RCA teams can vary in size, sometimes reaching 15 attendees or more. The space needs to be large enough to accommodate the team. It is also important the room be arranged in a way that supports good communication, and for the team to focus on the display of the chronology, Logic Tree, and action plan. We will say more about facilitation of the meeting itself in Chapter 6.

A final logistical preparation is to think about the materials needed. These might include a sign-in sheet, confidentiality agreements, post-it notes, easel pads, pens, and possibly a projector or monitor for display. Some facilitators like to provide an agenda. Whether distributed or displayed, an agenda can help orient the RCA team to the process. For those who have not previously attended an RCA team meeting, it will help them know what to expect.

In our experience, successful RCA team meeting agendas are straightforward. They typically include an introduction, statement of confidentiality, review of the chronology, development of the Logic Tree for analysis, and then documentation of an action plan.

3.3 EXERCISE 1: CHRONOLOGY

Your team has been assigned to investigate and conduct an RCA after the cancellation of a surgery due to the lack of a necessary supply. A special mesh, preferred by the surgeon for certain difficult abdominal repairs, was not available in the operating room. Unfortunately, the patient had already been put to sleep. The case was canceled and rescheduled.

1. What steps will you take in preparation for the RCA team meeting?
2. Considering the following facts, develop a chronology to present to the RCA team.
   - Ms. Fortune initially saw her surgeon, Dr. Dogood, in the clinic. Dr. Dogood recommended surgery to revise an abdominal repair that was leaking and discussed possible risks and benefits of the procedure. Ms. Fortune agreed and the procedure was scheduled for the next week.

- Dr. Dogood asked her staff to complete the consent form and to schedule the surgery. The office nurse filled out the consent form for Dr. Dogood to have Ms. Fortune sign.
- The office clerk faxed a copy of the request for surgery time, along with the consent form, to the hospital scheduler.
- The scheduler added Ms. Fortune to the surgery schedule.
- The surgery charge nurse reviewed cases for the next day, including Ms. Fortune's.
- The day of surgery, the circulating nurse brought Ms. Fortune into the Operating Room and confirmed her name and procedure.
- Ms. Fortune was sedated by the anesthesiologist.
- The surgeon arrived in the room and led a final time out. She asked whether the special mesh was available.
- The team realized the supply was not available and contacted Sterile Processing.
- The case was canceled and rescheduled.

3.4 SUMMARY

Pre-work for an RCA team meeting is critical to success with the meeting. The information gathered, preparation of the chronology, development of the Top Box, and management of logistics set the stage for rigorous analysis and action planning by the RCA team.

3.5 QUESTIONS TO CONSIDER

1. What processes are in place or needed in your organization to conduct early investigation of safety events?
2. How are leaders and RCA facilitators prepared to practice humble inquiry?
3. What logistical arrangements can your organization put in place to support RCA team meetings?

NOTES

1. American Society for Healthcare Risk Management (ASHRM), *Serious Safety Events: A Focus on Harm Classification: Deviation in Care as Link Getting to Zero*. White Paper Serious-Edition No. 2 (Chicago, IL: ASHRM, 2014).
2. E. Schein, *Humble Inquiry: The Art of Asking Instead of Telling* (Oakland, CA: Berrett-Koehler Publishers, 2013).
3. The Joint Commission, "Sentinel Events," www.jointcommission.org/resources/patient-safety-topics/sentinel-event/.
4. National Quality Forum (NQF), *Serious Reportable Events in Healthcare-2011 Update: A Consensus Report* (Washington, DC: NQF, 2011).
5. Ibid.
6. TJC, "Sentinel Events."

4 Creating the Logic Tree

4.1 LOGIC TREE ROOT CAUSES

Once the Event and Mode have been identified in the Logic Tree Top Box, we have reached the stage in the RCA for the team to analyze how the unintended event happened. From the outcome, the unintended event, we begin to work our way backward to understand the sequence of events that led to that outcome. Imagine a recording of the event as it unfolded. The Logic Tree is a means of viewing this recording in reverse.

A Logic Tree contains three forms of causal roots below the Top Box; from top to bottom they are Physical Roots, Human Roots, and Latent Roots. (Figure 4.1 shows the definitions for the Logic Tree.)

Physical Roots are typically the consequences of decisions made in error. How do we know an error occurred? What was observed? These physical manifestations are what finally bring to light the evidence of the vulnerabilities in our systems of care. They are sometimes referred to as "the sharp end" because they are found at the point where the patient is impacted. This is the reason the Physical Roots are at the top of the Logic Tree.

To move beyond the Physical Roots we ask, *"How could?"* these Physical Roots occur? Working our way down in the Logic Tree structure, we find *Human Roots* are the decision errors made by caregivers that led to the physical consequences. The decision may be an error of either omission or commission. In other words, Human Roots are decisions made in error to act in an incorrect manner, or decisions made in error to not act when it was needed. Human Roots are decisions that led to not doing something a caregiver should have done or taking an action they should not have taken.

James Reason helps us understand the kinds of human errors caregivers make.[1] In the Logic Tree we are documenting as a Human Root the decision made in error at that point. (Human Roots are not the same as *Human Factors*, which is the domain of interactions between people and elements of a system, and failures in Human Factors is a category of *Latent Roots*.) When the team reaches the point of a decisional error, a Human Root, the question is *"Why?"* the caregiver, the decision maker, made that decision? What was the thought process? The answers to these questions lead us to the final level of causes, the Latent Root.

Latent Roots represent the organizational systems upon which decisions are made: they are policies, procedures, and protocols; they are practice guidelines and bundles; they are physical and technological systems; they are cultural norms. Think about the various flows in healthcare: patients, families, supplies, medications, information, and so on. Each of these flows has inherent systems. They underpin the decisions made by caregivers at any step along the way. Caregivers rely on these

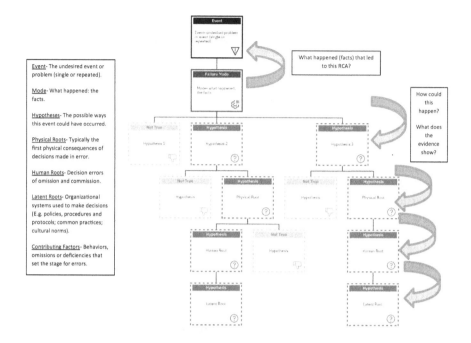

FIGURE 4.1   Logic Tree with Definitions

Source: EasyRCA image used by permission of Reliability Center, Inc.

systems to make decisions. Consider how the organization purchases equipment, supplies, and medications. What system is in place for managing purchasing? How are decisions made about what to purchase, and how to implement newly acquired equipment or supplies? How is inventory control managed? What system is used to determine which products are used first? How do we know they have not outdated? How are decisions about clinical practice decided and disseminated? What is the cultural norm about raising questions or safety concerns? Each of these contributes to the day-to-day decisions made by caregivers. To understand how a caregiver arrived at a particular decision in error, a Human Root, we need to look more deeply at the underlying systems, the *Latent Roots*.

One strength of the Logic Tree methodology is its drive to the Latent Root level. Too often in healthcare, there has been a tendency to stop at the Human Root level. Reason differentiates between the "person approach" and the "system approach."[2] If we stop at the Human Root level and blame the individual caregiver, "the person," for a mistake, we run the risk of not identifying how the decision error was made and fail to address the "the system" risk that will enable reoccurrence of the unintended event.

Ideally, our systems of care would be 100% reliable and would serve as the defense against decision errors. The Swiss cheese model[3] (Figure 4.2) illustrates how vulnerabilities in our systems, with varying levels of unreliability, can align and allow errors to travel.

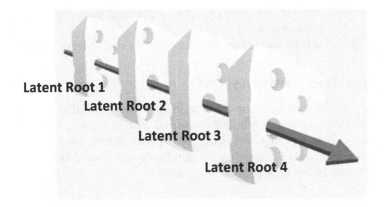

FIGURE 4.2   James Reason's "Swiss Cheese" Model

Sometimes the errors are prevented from reaching the patient, resulting in a Close Call or Good Catch. Sometimes they penetrate all our defenses, reaching the patient and causing harm. In both cases, the latent vulnerabilities in our systems become apparent. Reason describes latent conditions as "the inevitable 'resident pathogens' within the system. They arrive from decisions made by designers, builders, procedure writers, and top-level management."[4] These holes in our underlying systems influence decisions, which impact patients.

In summary, the Logic Tree allows us to work our way backward from the unintended event to understand the sequence of physical consequences of decisional errors based on underlying systems.

4.1.1 Exercise 2: Types of Roots

You have been assigned to investigate and conduct an RCA about a passenger train that derailed in the mountains between Chemult and Oakridge, Oregon. Miraculously, no passengers were hurt. The engine was a total loss. Interviews and data collection have revealed several facts. To begin, label the following as Physical, Human, or Latent roots:

- The most common cause of a derailment is problems with the track. An engineer has examined the track and did not find any breaks.
- The accident occurred on a curve, as the train was coming down the mountain. The engine was found lying on its side on the outside of the curve.
- The curve is marked at 45 mph.
- An examination of the "black box" reveals the speed registered deceleration from 70 mph to 50 mph prior to the accident.
- The conductor tells you the train was behind schedule that day.
- The train is often behind schedule because freight trains do not yield the right of way to the track.
- Examination of the brakes reveals less than 10% braking capacity remaining.

- A maintenance supervisor tells you this engine was three months overdue for routine maintenance.
- Funding cuts have caused re-prioritization of the maintenance schedule.

4.2 LOGIC TREE HYPOTHESIZING

A critical step in thorough and credible RCA is the generation of hypotheses. *How could* this healthcare error have occurred? In this section, we describe two approaches to hypothesizing. We share our experience with some of the risks in hypothesis generation that jeopardize the identification of latent root causes. And we identify choices for the RCA facilitator to consider.

4.2.1 HYPOTHESES

What is a hypothesis? It is an idea to be tested against evidence and is either provable or disprovable. The Logic Tree methodology requires asking the question, "*How could this error have occurred*?" Asking "*How could?*" leads to generating conceivable roots, starting with the Physical Roots. Let us look at some examples, starting with the non-healthcare event of Flight 173. Imagine how the investigators began reviewing the crash (see Chapter 1). When the NTSB investigators arrived, they would examine the physical wreckage of the airplane, along with data retrieved from the flight data recorder and the cockpit voice recording, as well as conducting interviews. Even prior to the results of these studies, they could create several hypotheses, such as: Flight 173 crashed because of catastrophic mechanical failure, or a problem with fuel, or a problem with navigation. All three elements are necessary for a successful landing: functional engines, fuel to the engines, and piloting to the runway. A failure in any one or more of these could have resulted in the crash. In fact, the NTSB investigators most likely generated these hypotheses even before their arrival in Portland at the crash scene. They could do so based on their knowledge of the causes that were found to have led to prior aviation accidents. Their task on arriving was to ask, "*What does the evidence show?*" These hypotheses can be shown in a Logic Tree. (See Figure 4.3 for an illustration.)

Think about healthcare failures of which you are knowledgeable, with readily available hypotheses. Take, for example, a patient fall. How could the patient have fallen? Based on evidence, we know patients can fall due to a physiological reason such as a syncopal episode, weakness, or dizziness. And we know patients can slip or trip on something in the environment. How could a medication error occur? It could be the result of an ordering error, a preparation error, or an administration error. We know this from experience with other medication errors.

These examples have illustrated hypotheses at the top of the Logic Tree. In fact, hypothesis generation and testing continue down the Logic Tree during identification of the Physical, Human, and Latent Roots. The discovery of a pressure injury provides an example. The first question is whether the wound was present on admission or healthcare acquired. (Figure 4.4 shows a Top Box with two initial hypotheses.) These are conceivable Physical Roots.

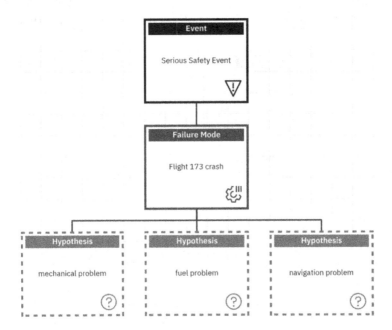

FIGURE 4.3   Flight 173 Top Box with Hypotheses

Source: EasyRCA image used by permission of Reliability Center, Inc.

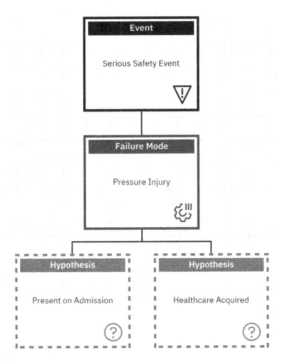

FIGURE 4.4   Pressure Injury Top Box with Two Hypotheses

Source: EasyRCA image used by permission of Reliability Center, Inc.

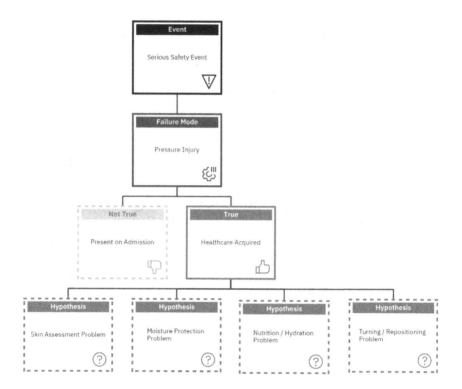

FIGURE 4.5   Pressure Injury Top Box with Continuing Hypotheses

Source: EasyRCA image used by permission of Reliability Center, Inc.

Once the evidence supported that the wound was not present on admission, the next layer of hypotheses can be investigated, which are usually Human Roots: representing the decisions made that caused the Physical Roots. Techniques for preventing pressure injuries include skin assessment, moisture protection, optimized nutrition and hydration, and turning/repositioning. These become the next level of hypotheses to investigate. (Figure 4.5 shows the addition of further hypotheses.) At this level, caregivers made either a decisional error of omission or commission. Evidence is used to confirm or deny the hypotheses. Any that are confirmed become a Human Root.

Thereafter, each Human Root is hypothesized for *how could* that decision have been made. Again, evidence is used to confirm or deny the hypotheses. Any that are confirmed become a Latent Root.

4.2.2 How Could versus Why

Notice the question in a Logic Tree is, *"How could?"* rather than *"Why?"* We have found two important reasons for this approach. First, asking *why* something happened or *why* a person made a particular decision is often experienced as casting blame. Despite our best intentions, it is a cultural norm that being asked *why* often results in people feeling defensive and contributes to a need to justify their actions. This response can inhibit our ability to understand the decisions made and, in turn, our ability to uncover the latent root causes of an unintended event. Second, asking *why* suggests there is one root cause. This is one of the limitations of the 5 Why methodology pointed out in Chapter 2. Asking *how could* an unintended event occur opens up the consideration of multiple possibilities. Rather than prematurely narrowing the analysis, this approach encourages investigation of more than one potential cause. The rigor and credibility of the RCA is strengthened if we can demonstrate that multiple possible causes were considered and ruled in or out based on evidence. In fact, we have found it is more common for RCAs to identify multiple latent root causes than a single path. Asking *how could* rather than *why* better facilitates their discovery.

4.2.3 Two Approaches to Hypothesizing

Think back to the two examples of hypothesizing discussed earlier: the fall and the medication error. In the case of the fall, we might generate hypotheses beginning with two known Physical Roots of falls (physiological and environmental). In the case of the medication error, we might generate hypotheses based on the steps in the medication process flow from the beginning step of manufacturing to the final step of monitoring the effectiveness of administered medication.

These examples represent two basic approaches to generating hypotheses. The first approach begins with an examination of the physical evidence when asking, *how could* a fall have happened. In the case of one fall, a hospitalized patient was found on the floor between the bed and the bathroom. When asked what happened, she reported to her caregivers that she was on her way to the bathroom when she slipped. This evidence, along with an assessment of her mentation, helped rule out the physiologic hypothesis and confirmed an environmental problem. It turned out the non-slip socks she wore were too large and had turned so the non-slip surface was not on the soles of her feet.

The second approach is useful when there is a defined process with identifiable steps. It is possible failures may have occurred at more than one step in the process and, therefore, important to look for evidence of failure at each step. In the case of one RCA related to an IV medication error, the hypothesizing began with each process step: the stocking and picking supplies from the supply room, the mixing of the medication in the clean room by a pharmacy technician, the labeling of the bag, the checking of the IV bag by a pharmacist, the delivery of the bag to the hospital unit, the confirmation of the patient and medication by the bedside nurse, and the programming of the pump and hanging the bag. Each step was hypothesized to have failed due to error. In this instance, a wrong supply was picked from the stock and used to mix the medication. That was two errors. A third error occurred when the pharmacist's double-check did not catch the incorrectly used supplies. The incorrectly mixed IV medication reached the patient and resulted in adverse symptoms before being identified. Applying hypotheses to each step in the process helped direct the team and identify the multiple errors.

4.2.4 Boolean Logic

The two strategies discussed earlier for hypothesizing suggest a form of logic needed, called Boolean logic, named for the mathematician George Bool. The logic is based on the concept that two hypotheses can be either conjunctive or disjunctive. In simpler terms, we might say hypotheses A *and* B are both true (conjunctive), or either A *or* B is true (disjunctive). In the case of the patient fall, we hypothesized that *either* the patient had a physiological event, *or* there was an environmental problem. (See Figure 4.6 illustrating the hypotheses and disjunctive logic.)

In the case of the IV medication error, evidence showed that an incorrect stock was picked to mix the IV medication, *and* the double-check by the pharmacist did not catch the mistake. (See Figure 4.7 illustrating the hypotheses and conjunctive logic.)

We will say more about the use of Boolean logic in Chapter 5 on telling the story of the RCA.

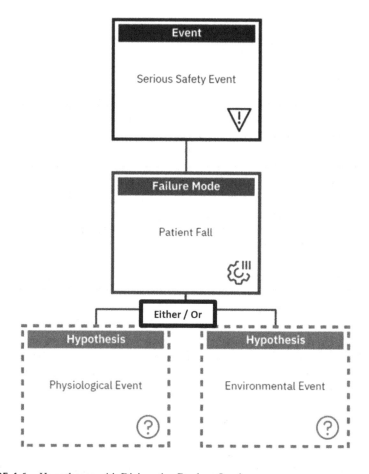

FIGURE 4.6   Hypotheses with Disjunctive Boolean Logic

Source: EasyRCA image used by permission of Reliability Center, Inc.

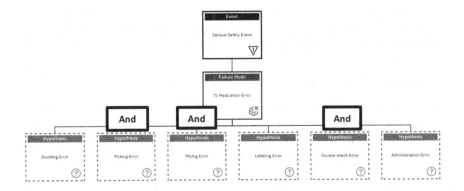

FIGURE 4.7 Hypotheses with Conjunctive Boolean Logic

Source: EasyRCA image used by permission of Reliability Center, Inc.

4.3 RISKS ASSOCIATED WITH HYPOTHESIZING

Over the years, we have made several mistakes in hypothesizing. Unfortunately, without well-stated hypotheses, the RCA is at risk of failing to clearly identify causal links leading to latent roots. Let us describe a few of the mistakes we have made and seen.

4.3.1 CATEGORIES VERSUS HYPOTHESES

Some accrediting organizations recommend or require RCAs to consider various categories, such as staffing, patient assessment, and care management. The categories have been derived from previously reported events. This requirement is intended to assure thorough RCAs. However, categories are not in themselves hypotheses (and are also not sufficiently comprehensive). The distinction between a category and a hypothesis begins with the way it is stated. A hypothesis statement framed as an error or failure could be, "patient assessment error." This differs from the statement, "patient assessment," which is not a hypothesis. Beginning with a set of categories at the hypothesis level is essentially making the Logic Tree into a Fishbone diagram. This approach does not lead to identification of Physical, Human, and Latent roots linked in causal chains.[5]

4.3.1.1 Countermeasure

One test for the adequacy of the hypothesis statement is how well it fits with the question, *"How could?"* For example, the question, *"How could* there be patient assessment?" doesn't make sense. Asking *"How could* there be a patient assessment error?" is a logical question that leads to the review of evidence and additional hypothesis generation. For example, the hypothesis statements "the patient was not assessed" and "the patient was assessed incorrectly" follow logically from "How could there be a patient assessment error?" Logic suggests *either* the patient was not assessed, *or* the patient was assessed erroneously. Evidence will be used to support one or the other of these hypotheses, leading to the next level of hypothesis generation, *"How could that happen?"* Our learning: frame your hypotheses as error or failure statements. You can test these with your evidence.

4.3.2 LATENT ROOTS AS HYPOTHESES

When we began using the Logic Tree with teams, we would display the Top Box and then ask the team, *"How could* this occur? The intent was to generate a list of hypotheses. What we have found repeatedly is that teams will often suggest a latent root as an initial hypothesis. In our IV medication event, for example, a team might suggest "lack of a reliable medication process" as a hypothesis statement. Reliable processes relate to our systems of policies, procedures, protocols, equipment, communication, and so forth. These *processes* or *systems* support the decisions made by caregivers. They are, in fact, our Latent Roots which become apparent when there is an unintended event.

4.3.2.1 Countermeasure

A test of the initial hypothesis statement is to ask whether, if confirmed, it is a Physical Root. Think about the two hypotheses in the patient fall event (physiological and environmental). If the patient had experienced a syncopal episode, confirming the physiological hypothesis, this is a Physical Root. For the patient who slipped, the confirmed environmental hypothesis is a Physical Root. Contrast this with a hypothesis that the medication error was caused by the "lack of a reliable process." This hypothesis does not describe a Physical Root, such as an error picking the right stock, an error in mixing the IV bag, or an error in the pharmacist double-check. Each of these describes a potential Physical Root.

The sequence of the Logic Tree progresses from first confirming the Physical Roots, to identifying the Human Roots that resulted in that physical outcome, and then to the systems, the Latent Roots, underlying those decisions. When generating hypotheses at the top of the tree, the hypothesis statements are almost always Physical Roots. In a small number of cases, we have found the Human Root led directly to the Mode.

4.3.3 HYPOTHESES ARE TOO BROAD

In one of David's first experiences using the Logic Tree, he asked the team to generate a list of hypotheses for how a patient could have eloped from a locked psychiatric unit. The team was aware of the basic facts of the case, including information that the patient had followed a visitor out the front door. Learning to use the Logic Tree and being cognizant of the recommendation to consider multiple hypotheses, David led the team in generating a long list from a Mode that just said, "Patient eloped." Now, a brainstormed list of the myriad of ways a patient *might* successfully elope could be useful for future safety enhancements (FMEA), but it didn't help bring clarity to how *this patient eloped.*

Part of the art of facilitating an RCA is helping the team to generate hypothesis statements based on the information obtained in the chronology of the case that can lead to the Mode. Had the Top Box in this case included a Mode Statement something like, "Patient eloped through front door of unit at 3:30 pm" it would have helped guide the hypotheses statements. It would lead to investigating how that specific Mode could happen, rather than any means by which a patient could elope. This approach causes investigation of the facts of the event and is more respectful of the precious resource of the team's time.

4.3.3.1 Countermeasure

As evidenced by this example, one strategy to reduce the risk of hypotheses being too broad is to carefully craft the statement in the Mode. Agreement from the team on *what* is being reviewed aids in appropriately narrowing the focus of *how* the event occurred. Equally important is to facilitate the RCA team using deductive reasoning. Hypotheses should be deduced from the known facts. For example, in the case of the eloped psychiatric patient, David might have prompted the team to hypothesize how the patient was able to follow a visitor out the door. Hypotheses might have included inattention by staff, lack of an anteroom, poor visualization of the door, etc. These more accurately fit the facts and would lead to meaningful action plans.

4.3.4 HYPOTHESES ARE TOO NARROW

The opposite of the problem of generating hypotheses not based on the known facts and the Mode is the temptation by teams to limit hypotheses based on the belief they know how an unintended event occurred. The stage of initial hypothesis generation is one most at risk of the influence of bias. We all have our cognitive biases. Purposefully asking a team to *logically* consider all the ways in which an unintended event could have occurred is an important step to allow the team to help overcome individual bias.

Hierarchy and the power gradient play a role here. Physicians are enculturated to be the "captain of the ship." Medical training includes residents summarizing a case and reaching a diagnosis and treatment. Some physicians may approach RCA in the same way. Given the relative level of training, power, and deference to physicians, a team may simply accept the physician's understanding of how an unintended event occurred and not consider additional hypotheses. The RCA facilitator may need to support the team in thinking more broadly about possible hypotheses. For example, an RCA conducted by Harold and David focused on a wrong-level spine surgery. Initially, the spine surgeon wanted to accept responsibility, and to share an assessment of individual fallibility as the cause of the unintended event. By proceeding with the Logic Tree approach and generating multiple hypotheses, it was discovered that the event was a result of a combination of Latent Roots. These included the fact that the spine surgeon counted vertebrae from the sacrum up for some cases while the radiologist always counted and marked films (when film was still used) from the top. This wasn't discovered until well into the RCA with the participation of both the spine surgeon and radiologist. Had the team accepted the spine surgeon's sole hypothesis that he was at fault, they would not have reached this important conclusion allowing a change in practice and improved patient safety. The team might not have looked beyond the Human Root level (the spine surgeon erred) to the Latent Roots.

4.3.4.1 Countermeasure

Strategies to address the risk of narrow hypotheses and unconscious bias, include the assembling of a multidisciplinary team and recognizing how each caregiver was involved in the case. This tends to bring forward the complexity and interplay of multidisciplinary care. It is important to invite all members of the team to help generate the initial list of hypotheses. One of our colleagues introduced a practice we have adopted: leaving a blank box at the top level of hypotheses. This signals the team that there may be additional hypotheses to be discovered and considered as the RCA proceeds. It communicates permission to add to the list. It is also important to recognize that we, as RCA facilitators, have our own cognitive biases. By conducting interviews and thorough records reviews, we may believe we know *how* this unintended event occurred. We are in danger of coming into the RCA team meeting and influencing the team based on our pre-assessment. In fact, David and Harold have experienced this many times. In several cases, it was not until the team met and reviewed the Logic Tree together that important additional evidence emerged. Our preconceived ideas about the causation of the case turned out to be wrong. While we can never completely escape our biases, we can take steps to hold them in check. Facilitating everyone's engagement and mindfully remaining open to new information as it emerges in the RCA are vital.

4.4 THE IMPORTANCE OF FACILITATING HYPOTHESIZING

As we shared, the generation of good hypotheses at the top of the Logic Tree is a key to RCA success. Skillful facilitation is required, and facilitators should consider and plan their approach. Two strategies are helpful: thinking about hypotheses in advance of the RCA team meeting and deciding whether or not to pre-populate Physical Root hypothesis boxes.

First, it is natural for RCA facilitators using the Logic Tree methodology to begin formulating hypotheses while conducting early investigations. We find ourselves in the mindset of asking, *"How could?"* and developing hypotheses as we continue to gather evidence. If the facilitator is experienced with unintended events and healthcare errors, hypotheses based on prior RCAs should be considered. Staying mindful that you may be missing information, and emphasizing your role facilitating rather than imposing expertise, you can develop reasonable proposals for hypotheses to share with the RCA team.

Second, another strategic decision is whether to pre-populate the first (top) level on the Logic Tree with Physical Root hypotheses or to go into the RCA team meeting with nothing pre-populated. In some cases, you, the RCA facilitator, may have gleaned information from your interviews and early investigation that can be called out as hypotheses. For example, while investigating a case of a retained broken peripheral IV catheter tip, David learned of several risks of failure. Failures could have occurred at the point of placing the IV, while maintaining the IV, and

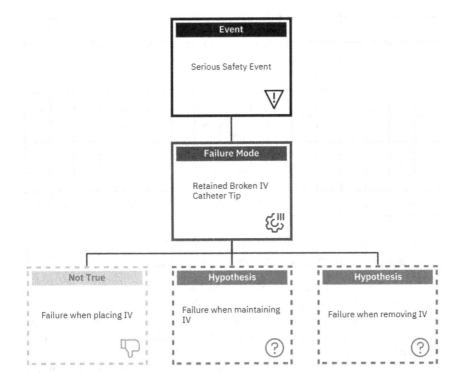

FIGURE 4.8   Retained Catheter Tip RCA with Three Hypotheses

Source: EasyRCA image used by permission of Reliability Center, Inc.

at the point of discontinuing the IV. (See Figure 4.8 illustrating the Logic Tree with Physical Root hypotheses.)

These translated into three initial Physical Root hypotheses for how the patient could have suffered a retained broken catheter tip. These were pre-populated and displayed on the Logic Tree, and then confirmed by the RCA team. The team was able to quickly rule out, based on evidence, catheter failure at the time of placement. They turned their attention then to evidence of how the catheter could have failed during maintenance and/or when discontinuing.

There are advantages to pre-populating hypotheses on the Logic Tree. It can help the RCA team understand the use of deductive reasoning. It can guide the team away from jumping to latent roots during the initial hypotheses, and toward the proper sequence of identifying Physical, Human, and Latent roots. It provides recognition for the information gathered in advance and it can be an efficient use of time.

On the other hand, there are risks. We run the risk of introducing the bias of those interviewed, or our own. We can infer these are the only hypotheses and shut down exploration of alternatives. We have found that decisions about pre-populating the initial Physical Root hypotheses are best made on a case-by-case basis. Considerations include the team composition, the complexity of the case, the complexity of the logic of the initial hypotheses, and the skill level of the facilitator.

4.4.1 EXERCISE 3: HYPOTHESIS GENERATION

Consider these scenarios and based on your experience and knowledge of the processes and potential failures, generate a list of potential, starting hypotheses.

1. You are asked to lead an RCA involving a contaminated instrument on the back table in a surgical case. The instrument is discovered as the back table is being opened, prior to the arrival of the patient. The instrument contains bioburden. How could this happen?[6]
2. A primary care provider referred a patient a year ago to a specialist. On the patient's return to the primary care provider for their next annual visit, it was discovered the patient never saw the specialist. You are asked to lead an RCA and determine how this failed referral occurred. How could this have happened?[7]

4.5 VERIFYING AND TESTING THE LOGIC TREE

By now, we have defined the Top Box of the Logic Tree, as well as the Physical, Human, and Latent Roots. We have described the importance of generating initial Physical Root hypotheses, followed by hypotheses for the Human and Latent Roots. We turn now to the use of verifications and how we test the logic of the Logic Tree to confirm the chain of causation. This is how we can be confident we have discovered the Latent Root causes of an unintended event. If we fail to identify the Latent Roots, we are at risk of repeating the unintended event.

4.5.1 VERIFICATIONS

The two seminal questions of the Logic Tree are: *How could this occur*, and *What does the evidence show?* Each hypothesis needs to be tested with the evidence to either support the hypothesis as a true statement or rule it out. In Chapter 3, we went into the practices for gathering information in the early investigation phase of an RCA. Here we focus on the types of evidence we might use for verifications, and how they are used within the Logic Tree.

What kinds of evidence are used in healthcare RCAs? Let's look at some of the sources of data. Two of the most readily accessible sources of evidence are the caregivers involved in a case and medical records. An advantage of having an immediate response system is the ability to gather information through interviews and observation with those present at the time and at the location of an unintended event. Seeing the scene and environment can help with building an understanding of the facts. Numerous healthcare errors occur in the interface between caregivers and technology. The ability to visualize (and possibly photograph) patient rooms and equipment can be important to understanding the information from interviews. Typically, additional caregivers will need to be interviewed either because they have left the area or because the event was discovered after the fact. The RCA facilitator can develop a list of key interviews to conduct prior to the RCA team meeting. Interviews are a primary source for obtaining facts: what happened, in what sequence, who was involved. Most importantly, interviews are the opportunity to understand what the involved caregivers knew at the time of the event. Memories change with time and from the bias of knowing the outcome. Understanding the knowledge base and the state of mind of the caregivers is critically important to later documenting the Physical and Human Roots of the unintended event.

A second common source of evidence with which to verify hypotheses is the medical record. The Electronic Medical Record (EMR) contains a large volume of data, and it is often presented in a view unique to a caregiver's role. The view of a physician may differ from that of the nurse, and the therapist. Sometimes there is customization to the individual caregiver level. Considering this, it is important when retrieving data from the EMR to be mindful of the roles and views of those involved. It may be helpful to have a nurse manager, medical director, analyst, or informaticist help locate, retrieve, and interpret the patient's record.

While interviews of the people involved and the review of the EMR are two of the greatest sources of data for the RCA, there are many more to consider. Latino[8] suggests we look to 5Ps: parts, position, people, paper, and paradigms.

4.5.1.1 Parts

Did the unintended event involve any equipment? Think about what is in a hospital room: there is usually a bed or stretcher, a headwall with gas connections, monitors, and pumps. Could any of these have played a role in the event? Is it possible to retain the piece of equipment for interrogation? If not, can photographs be made to document the type of equipment and its state at the time of the event? In surgical cases, it may be challenging to examine a non-sterile instrument, but can it be safety preserved? Can a duplicate of the same instrument be obtained for the RCA team to examine? Medication bags should be retained. Photographs of the labels on the bag may be valuable. These are examples of the kinds of parts that provide evidence for an RCA.

There is often a digital "footprint" connecting parts in the hospital or clinic. Think, for example, about a medication error involving a dispensing machine. These "smart" pieces of equipment may be connected electronically, creating a digital record with time stamps. Reports may be obtained from medication dispensing systems, nurse call systems, bed tracking systems, and more. Consider the records associated with any part of interest.

4.5.1.2 Position

In some healthcare errors, the importance of position is readily apparent. Take, for example, the case of a wrong-side surgical procedure. The RCA team will need to understand the position of the patient on the table or procedural chair, and the position of each caregiver relative to the patient. Position may also involve the movement of caregivers, supplies, or information. Think about how to convey this data to the RCA team. Photographs, drawings of the room, and floor plans may all help the team to visualize positioning of people, supplies, and information. Facilities departments often have floor plans. In one case, David used a floor plan with an overlay of data collected from the nurse call system to create a spaghetti diagram of caregiver movement. This was critical to understanding the response time to a patient who fell. In a retrospective analysis of a tragic MRI error resulting in the death of a pediatric patient, Latino and Gilk[9] use animation to show the movements of those involved within the MRI suite. This display of positioning of the anesthesiologist, the MRI techs, a nurse, and the code team facilitated the analysis of the case using the Logic Tree.

4.5.1.3 People

We described the importance of interviewing caregivers directly involved with an unintended event. Are there caregivers upstream from the event who may have additional information? These might include caregivers from prior shifts, or previous patient care locations. They can also include charge nurses, supervisors, managers, clinical educators, and informaticists, among others. In cases where the discovery of the event occurs without the patient being present, often true of the ambulatory setting, the RCA facilitator may need to dig into the case to discover the people with valuable information. Compounding this is the fact that many of the serious safety events in the ambulatory setting involve failures of intended action. These may be errors of omission rather than commission. Identifying the intention of the clinicians, and the normal process, may be necessary to identify caregivers to interview.

4.5.1.4 Paper

Most health records are now electronic, but they can still be thought of in the category of paper. What other kinds of documents might be useful? Consider both electronic and paper versions of documents. They might include policies, procedures, and protocols; training and competency records; equipment maintenance records; signage; supply and equipment purchase histories; floor plans, and more.

4.5.1.5 Paradigms

A paradigm is a model. What kind of data might be gathered from paradigms? In our experience, it is helpful to understand the culture of the organization, unit, or department, and shift involved in an unintended event. Culture is driven by the norms for behaviors, the language used, the acknowledgment of important events, and the level of engagement. These factors may not be documented, per se, but are felt and experienced by caregivers. The unwritten rules often guide decisions and actions: they are often the Latent Roots.

These paradigms can be assessed through interviews with caregivers. Insights into daily decisions are gleaned with questions such as, "What normally happens?", and "What do caregivers on other shifts or in other departments do?", "How is it different here?" An important part of a just culture algorithm is the "substitution test." That is, if one caregiver decides to act one way, what would others decide? This can be helpful in understanding Human Roots in a Logic Tree.

As may be clear from this discussion of the types of data you might collect, numerous sources of information help develop a chronology of a case and support the verification of hypotheses in a Logic Tree.

4.5.2 Documenting Verifications

Through data gathering in the early investigation phase, and by assembling a knowledgeable team, evidence is used to guide the RCA team down the Logic Tree. It is important to verify in the Logic Tree the decisions to rule hypotheses in or out. This is a key feature distinguishing the Logic Tree from shallow forms of RCA. The documentation can be supported by software functionality or by paper documentation. In either case, it supports the conclusions reached in the case.

It is important to verify decisions to rule out hypotheses. For example, "inadequate staffing" may be included as a hypothesis in a case. Regulators are particularly interested in an association between staffing levels and unintended events. Through the review of expected staffing patterns and the staff on duty at the time of an event, the RCA team can verify whether the hypothesis is supported or not. Copies of the staffing records, embedded in the Logic Tree or kept in the RCA files, serve as the evidence of the verification. For similar reasons, verification of competencies, privileges, protocols, and so forth can be important support for the conclusions of the RCA team. (Figure 4.9 shows a verification within the EasyRCA Logic Tree software; Table 4.1 illustrates use of a table for verifications.)

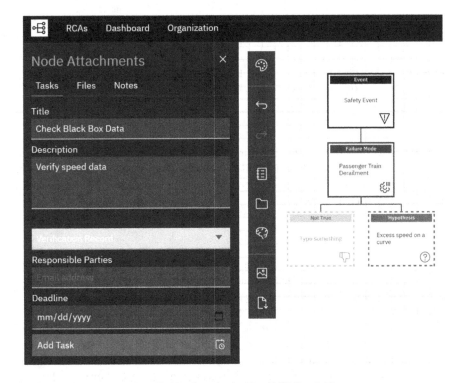

FIGURE 4.9 Documenting Verifications in the EasyRCA Logic Tree

Source: EasyRCA image used by permission of Reliability Center, Inc.

TABLE 4.1

Documenting Verifications with a List or Table

Resource Name	Hypothesis	Verification	Outcome
David Allison	Break in Track	Engineering inspection of track	No breaks in track found
David Allison	Freight Train Delay: Ongoing schedule delay for passenger trains	Management interviews and review of documents	Interviews and documents indicate priority given to freight trains over passenger trains
David Allison	Re-prioritization: Funding cuts cause re-prioritization	Management interviews and document review	Interviews with managers and review of meeting minutes and budget documents reveal re-prioritization of maintenance schedule
David Allison	Routine maintenance was out of schedule	Review of maintenance records	Records indicate engine three months past due for scheduled preventive maintenance
David Allison	Brake pads were low	Engineering inspection of brakes	Inspection reveals critically low level of braking capacity
David Allison	Train greater than recommended speed	Examination of "Black Box" data	Data shows speed slowing from 70 to 50 prior to derailment

A trap that some RCA teams fall into is a reliance on opinions about "what normally happens." Information on normal procedures, or expected processes, may be useful for generating hypotheses. However, the verification needs to be based on the evidence of what occurred *in this case*.

4.5.3 Strength of the Evidence

In healthcare, as in much of life, the strength of the available evidence varies. Healthcare delivery, patient factors, and pathophysiology involved in unintended events are complex. This leads to a range of evidence for verification.

Sometimes there is strong evidence available, as in the case of the patient who reported she slipped on her way to the bathroom. The evidence in this case was clear: the non-slip surfaces of her socks were out of position due to their large size. The combination of evidence available from the patient and from the environment allowed the team to reach strong conclusions as to the sequence of events. The team learned there were only two sizes of non-slip socks available in the inventory on the unit. They were then able to investigate how there were only two sizes for the caregiver to choose from. The Physical and Human Roots were evident, allowing the RCA team to draw strong conclusions on the Latent Root causes. In this case, the Latent Roots included purchasing decisions and limitations on supply distribution.

Other times the evidence will be less available or not as conclusive. In unintended events where the evidence is dependent upon pathology results, there can be a delay in the ability to verify. In some cases, such as when a cause of death is unclear and no autopsy is performed, needed evidence may never be obtained. Or the evidence may cause a delay in verifying hypotheses and reaching final conclusions. In one case, a surgical team identified a missing sponge following an abdominal surgery. The team was never able to locate the missing sponge. Two views were taken by X-ray and the surgical suite was searched but without success. Given the procedures followed after identifying the discrepancy in the count, it was unlikely the sponge was retained in the patient. However, physical evidence could not be used to verify the conclusion. In one case of a patient who died by gun violence, it was unclear whether the event was an accidental or intentional discharge of the gun. Autopsy results provided evidence in support of the wound being self-inflicted but could not address the question of intent.

In cases where it is unclear what happened in the patient's course, we may ask the physicians to provide an opinion based on their medical judgment. Is one hypothesis more likely than another? Is it more likely than not? That is, is there a greater than 50% probability the hypothesis is true?

The number of verifications for an RCA will vary by the complexity and size of the case. The evidence and the number of verifications needed correspond to the number of hypotheses. In some cases, verifications will be based on interviews, patient records, and possibly some policies and procedures. However, some cases may require more extensive evidence and documentation of verifications. David was asked by an executive to conduct an Operational RCA investigating deficiencies in the surge capacity of an emergency department. The Mode was based on caregiver safety concerns expressed one specific weekend. Three initial hypotheses were explored: demand exceeded capacity (H1); problems related to boarding patients (H2); and the surge plan was insufficient or ineffective (H3). A substantial investigation was required to understand the problems that culminated in the emergency department that weekend. The evidence gathered included interviews with multiple caregivers; reviews of patient records; examination of historic census data and trends, staffing patterns, productivity reports, door-to-bed data, dashboards, surge plans; and review of the policy and procedure for diversion. The Logic Tree resulted in the identification of 22 Latent Roots. Thirty-four records of verification were used to support the conclusions of the Physical, Human, and Latent Roots.

4.5.4 CONFIDENCE LEVELS

The strength of the evidence for a hypothesis can be rated as a confidence level. The PROACT software includes a confidence scale for roots. (Figure 4.10 shows documentation of confidence levels on Logic Tree boxes.)

A confidence scale may be added to other commonly used software or with paper-based versions. We have found it important to include the RCA team in assessing the strength of the evidence and the confidence level in the root cause. The scale

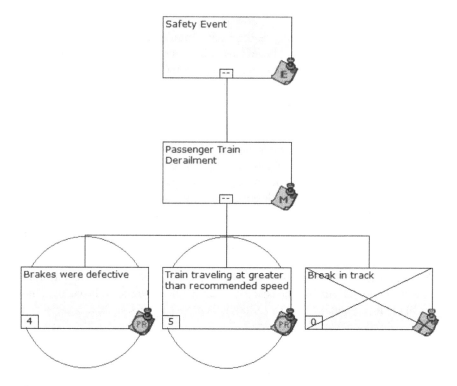

FIGURE 4.10   Displaying Confidence Levels

Source: PROACT image used by permission of Reliability Center, Inc.

consists of 0–5. A zero (0) indicates evidence the hypothesis statement is not true: it has been ruled out based on the evidence. A confidence level of 5 indicates the highest level of evidence is available to support that the hypothesis is true. A level of 1 means the hypothesis is hypothetically true but without evidence to verify. A confidence level of 3 is consistent with a 50% probability based on the evidence the hypothesis is true.

Let's look at a couple of examples. Take the case of the patient fall with the non-slip socks again. The physical evidence was strongly verifiable (confidence level 5) that the patient slipped, and the non-slip socks were not properly positioned. The decision by the nurse to provide the incorrect size of socks was verifiable at the highest level. The lack of availability of the appropriate size of non-slip socks in the unit supply was equally as verifiable. The Latent Roots of supply chain purchasing decisions without clinical input, and the limited unit storage space reducing the number of sizes of socks to choose from, were verifiable at the confidence level 5. On the other hand, in the case of the missing sponge after an abdominal surgery, the Physical Root could not be verified with confidence as the sponge was never located. The hypothesis that the sponge was retained in the patient was deemed probably untrue based on the evidence of the X-rays and examination of the cavity by the surgeon. However, it was theoretically possible the sponge was retained. The hypothesis could not be ruled out as untrue, so was assigned a confidence level of 1.

Most Logic Trees find multiple Latent Root causes for unintended events. The strength of the evidence for verification may vary among the Latent Roots. Some may be strongly verified by evidence and rate a high degree of confidence. Others may be more difficult to verify. These can include Latent Root causes describing culture, and other less tangible types of evidence. Documenting the evidence used to verify each root cause and assigning the appropriate confidence level to the root strengthens the conclusions of the RCA. It can also help in telling the story of the event.

4.5.5 CHAIN OF CAUSATION

We previously referred to the deductive reasoning used to construct the Logic Tree as playing the video in reverse. The Logic Tree begins with the Top Box and asks, *"How could this happen?"* Analysis progresses down the Logic Tree based on what the evidence shows, from Physical Roots to Human Roots and finally to Latent Roots. A Logic Tree may include multiple proven hypotheses at each level, depending on the complexity of the unintended event. Each verified root is causally linked to the next in the progression to the Latent Roots. The physical consequences were the result of the decisions made, which, in turn, were caused by the Latent Roots. (Figure 4.11 illustrates the chain of causation.) These causal connections make up the *chain of causation.*

The *chain of causation* provides a test of the deductive reasoning and confirmation of the RCA conclusions. You can confirm the logic by reversing the questioning. Starting at the bottom of the Logic Tree with the Latent Roots, you can trace the links up and ask, "Did this root cause the root above?" Using our patient fall case as an example, we can test the causal links by asking, did the purchasing decision (limiting supplies to two sizes) cause there to be only two sizes of non-slip sock in the unit supply? Yes. Did the limited sizes in the unit supply cause the nurse to pick an incorrect size for the patient? Yes. Did the incorrect size of non-slip sock cause the patient to slip and fall? Yes.

4.5.6 EXERCISE 4: LOGIC TREE

Using the facts from the train derailment exercise in Chapter 4, construct a Logic Tree with the following characteristics:

- Label the Event: "Logic Tree Training Exercise #1."
- The Mode is "Passenger train derailed."
- You have two hypotheses.
- You can rule out one of the hypotheses based on evidence.
- There is evidence to support the second hypothesis, including Physical, Human, and Latent Roots.
- Test the chain of causation for your Logic Tree.

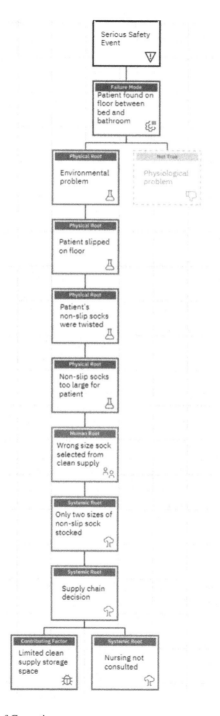

FIGURE 4.11   Chain of Causation

Source: EasyRCA image used by permission of Reliability Center, Inc.

4.6 LATENT ROOTS VERSUS CONTRIBUTING FACTORS

At this point, we need to distinguish between Latent Roots and Contributing Factors. Some of the underlying systems involved in a case may have set the stage for errors to occur but did not directly cause the error. We label these as Contributing Factors. (See Figure 4.1 for definitions of both Latent Roots and Contributing Factors.) David Gluzman[10] provides the following definitions:

> *Cause* is a condition that produces an effect; eliminating a cause(s) will eliminate the effect.

> *Contributing Factor* is a condition that influences the effect by increasing its likelihood, accelerating the effect in time, affecting severity of the consequences, etc.; eliminating a contributing factor(s) won't eliminate the effect.

The distinction is important as we develop action plans. To effectively eliminate the risk of a repeat unintended event we need to target 100% elimination of a Latent Root Cause.

It is common for RCA teams to identify conditions that may have contributed to an event without directly causing it. The chain of causation is the way to test whether a box at the bottom level of the Logic Tree is a Latent Root or a Contributing Factor. In our patient fall example discussed earlier, the limited space in the unit supply area did not directly cause the decision to purchase only two sizes of non-slip socks. It may have contributed to the decision to limit the number of sizes being stocked and is, therefore, a Contributing Factor. However, a larger storage space does not guarantee that additional sizes of non-slip socks will be purchased.

4.7 TELLING THE STORY

Communicating the conclusions of an RCA is important to organizational learnings and the prevention of reoccurring unintended events. Understanding the deductive reasoning of the Logic Tree and the chain of causation allows the story to be told. As the RCA facilitator, you can create a narrative beginning at the Latent Root level and explaining the causal linkages up through the Logic Tree. From our previous example, you might say, "Decisions made by supply chain to purchase only two sizes of non-slip socks, perhaps based in part on the limited storage space available on units, led to a nurse choosing a non-slip sock that was too large for the patient. The patient was wearing the non-slip sock, but it twisted on her foot and the non-slip portion was not on the bottom of her feet. As she was walking from the bed to the bathroom, she slipped on the floor in her stocking feet. As a result, she suffered a fall with injury."

Your Logic Tree is unlikely to show a single path from the Mode to the Latent Root. Remember, the strength of the Logic Tree methodology as compared to the 5 Whys methodology is the identification of the multiple Latent Roots. This means that telling the story of an RCA most likely involves the conjunctive, "and." This *and* that happened, leading to the undesired outcome. Boolean logic looks at the conjunctive or disjunctive nature of two hypotheses. When narrating how an unintended event occurred, we can speak to the combination of Latent Roots that ultimately led to decision errors resulting in physical consequences, all adding up to the unintended event.

4.8 SUMMARY

The Logic Tree methodology provides thorough and credible analysis, founded on evidence. Guiding the RCA team through inquiring *how could* the unintended event have occurred will successfully lead to identifying the multiple Latent Root Causes and Contributing Factors.

4.9 QUESTIONS TO CONSIDER

1. Is there a recently completed RCA from which you could try to construct a Logic Tree as described in this chapter? If so, what did that exercise reveal?
2. Are you using categories or hypotheses in your RCAs?
3. What method of collecting evidence does your organization currently use?
4. Is there opportunity to increase the rigor of your evidence gathering? What is the first step?
5. How will you document the evidence used in your RCAs?

NOTES

1. J. Reason, "Human Error: Models and Management," *British Medical Journal* 320, no. 7237 (March 18, 2000): 768–70.
2. Ibid., 768.
3. Ibid., 769.
4. Ibid.
5. See the discussion of the differences between Fishbone Diagrams and Logic Tree Analysis in Chapter Two.
6. Possible starting hypotheses include: the instrument was not properly processed in the operating room after the last case; the instrument was not washed; the instrument was improperly sterilized; the inspection process failed.
7. Possible starting hypotheses include: the specialist failed to receive the referral; the specialist failed to act on the referral; the patient chose not to follow through.
8. R. Latino, *Patient Safety: The PROACT® Root Cause Analysis Approach* (Boca Raton, FL: CRC Press, Taylor & Francis, 2009).
9. See "YouTube Columbini MRI Case Root Cause Analysis," accessed October 5, 2020.
10. D. Gluzman, "Cause vs. Contributing Factor," 2020, https://reliabilityweb.com/articles/entry/Cause_vs._Contributing_Factor.

5 Effective Action Plans

RCA[2] (RCA "squared") is a term advocated by the authors of *RCA2: Improving Root Cause Analyses and Actions to Prevent Harm*[1] as a counter to the common problem of healthcare RCAs not resulting in the implementation of effective action plans. Without improvement work to prevent future reoccurrence of serious safety events, the goal of RCA is not achieved. We have defined RCA as *a rigorous, standardized process designed to lead to tangible results for improving safety.* This requires both analysis and implementation of improvement strategies. This chapter will confront one of the most frequent failures of healthcare RCA: the failure to implement effective action plans. We will describe the novel approach we have taken to action planning, share an action plan rigor test, and discuss appropriate metrics.

5.1 THE REALITY OF MOST RCA ACTION PLANS

One of David's first assignments after moving from Clinical Risk Management into a newly created position dedicated to patient safety was to review the results of the prior year's RCA action plans. An analysis of ten RCAs conducted was done. The look back included going to the involved units to observe improvements and interviewing the unit leaders. Credible work done by RCA teams to analyze the causes of serious safety events was evident. However, the action plans mostly lacked the rigor needed to ensure patient safety. Further, their implementation was mixed. Of the ten cases reviewed, only two had strong action plans[2], six were intermediate in strength, and two were weak. (See Table 5.1 for definitions of action plan strengths.)

In addition, only two of the cases had fully completed action plans, six were partially completed, and two were delayed. This is not unusual across healthcare today; it is one of the reasons for the persistent level of patient harm.

5.2 A NEW AND NOVEL APPROACH TO RCA ACTION PLANNING

How then to address this reality as the leaders for RCA in our organizations? We advocate taking a new approach to action plans. Returning to David's retrospective review of the ten RCAs, there were several reasons that eight of the ten action plans were not fully implemented; three stand out. The first reason was the number of items included on the plans. This varied, but all of the plans had multiple recommendations for improvements. The time needed to carry through on the multifaceted action plans was significant. The responsibility often fell to the unit manager who had multiple competing priorities. The second reason was the degree of difficulty of the incomplete action items. Easier recommendations were often completed, but those more difficult to implement were sometimes left undone. This was usually related to the degree of difficulty to improve the latent root. Instituting strong controls to reduce

TABLE 5.1
Strong, Intermediate, Weaker Action Plans Defined

Action	Primary Analysis Category
Stronger Actions	• Architectural/physical plan changes • New devices with usability testing before purchasing • Engineering control, interlock, forcing functions • Simplification of processes and removal of unnecessary steps • Standardization of equipment or process or care maps • Tangible involvement and action by leadership in support of patient safety
Intermediate Actions	• Redundancy/backup systems • Increase in staffing/decrease in workload • Software enhancements/modifications • Elimination/reduction of distractions • Checklist/cognitive aids • Elimination of look- and sound-alikes • Enhanced documentation/communication
Weaker Actions	• Double checks • Warnings and labels • New procedures/memorandums/policies • Training • Additional study/analysis

the vulnerability for latent systems were often constrained by budget, limited information technology resources, or differing organizational priorities. The third reason was the lack of oversight. Leaders lacked a management system to track the completion of action plans, or to identify barriers preventing their completion. Incomplete action plans were not made visible.

To address the number of items included on action plans and their varying degrees of difficulty, let us to return to James Reason's Swiss cheese model of system accidents[3] (Figure 4.2). There are holes of different size in the Swiss cheese slices: the metaphorical holes of healthcare unreliability are caused by latent conditions. These are based on the degree of reliability of the system. The degree of unreliability, how often a system fails, corresponds to the size of the hole in the slice of cheese. Most systems have some degree of latent unreliability. That is, the likelihood is there will be some kind of failure. Mistake proofing, redundancy, inspections, and discoverability are some techniques we can build into our systems to provide a defense against a mistake or failure in one system or process step (slice of cheese). Unintended events occur when the holes in our systems align, and all our safeguards are penetrated. Remember that each slice of the cheese represents a system intended to safely support patient care but has some latent level of unreliability.

A well-done RCA identifies the holes in the systems and process steps, the latent roots, which aligned to result in an unintended event. Most RCAs, in fact, identify multiple latent roots. We have not conducted an RCA to date with only one Latent Root found.

Now think about each latent root as a slice of the Swiss cheese. Harold demonstrated this for a team by asking several volunteers to stand in a row and to each hold a piece of paper provided. He had cut holes of various sizes into each piece of paper to represent the varying unreliability of the latent roots. It was easy to see as the volunteers moved their individual pieces of paper around (reflecting daily variation) how there were times when all the holes aligned. Latent Roots, in a Logic Tree, function as Boolean AND operators: all are required for the failure Mode to occur.

Now imagine how the substitution of one complete piece of paper, or one slice of solid Cheddar cheese, eliminates the possibility of an error passing all the way through to the patient: the Boolean AND logic for causing the Mode is not met. Eliminating the vulnerability in any one of the latent roots will result in the elimination of the same Mode in the Top Box from reoccurring.

With this insight, we have begun taking a novel approach to creating an action plan after completion of the Logic Tree. As the RCA facilitator, we ask the team to confirm whether each root at the bottom of the tree is either a latent root or contributing factor. You can do so by asking, "If we eliminate this root, will it prevent this event from reoccurring?" By walking up the chain of causation from the roots at the bottom of the tree, you can confirm either they were causally linked, or a condition that contributed but did not directly cause the error.

Once the latent roots are confirmed by the team, the next step is to ask which of the latent roots can be eliminated most reliably. It is not necessary to improve each of the latent root causes, but only to eliminate one to prevent the same Mode of unintended event from reoccurring. (Figure 5.1 includes the steps for the RCA team.)

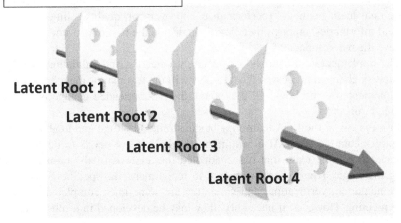

FIGURE 5.1 Swiss Cheese Model with RCA Team Questions

This can be challenging for an RCA team to determine. Having clinical content and improvement experts participate in this discussion is vital. It is worth taking the time to hear different perspectives and strive for a consensus. In the case of a broken peripheral IV catheter tip, the team identified multiple latent roots. They found vulnerabilities in the manufacturer's product defect rate, in the degradation of the tubing material due to practices of flushing with a small-bore syringe, as well as the length of time the catheter is in. Each of these latent roots: manufacturing defect, defects in practices for how the catheters are flushed, and variability in how long catheters are used were considered for elimination. The team determined impacting the manufacturing failure rate would be challenging. It was easier to imagine creating 100% reliability in the practices of flushing peripheral IV lines with the correct syringe size or creating 100% reliable practices for removing peripheral IV catheters based on best practice guidelines. Ultimately, the team opted to institute a daily review of peripheral IV catheters to support 100% timely discontinuation.

5.3 ELEMENTS OF THE ACTION PLAN

Using this approach to action plans, the RCA team is guided to select the latent root cause which in their determination will be most easily eliminated. The team is then tasked with recommending an action plan, which is predicted to be 100% successful at eliminating the latent root cause. Ask, "If we fully implement this recommendation, will it prevent this same Mode of unintended event happening again?" The team is challenged to form a recommendation that meets this test.

Once the recommendation, *what* to achieve, is decided on, the team then turns to *how*. What is the process for achieving this 100% reliable outcome? This is where the expertise in improvement strategies is helpful. Is an improvement team (Rapid Process Improvement or Kaizen) needed? Does an FMEA need to be conducted on a new process? Are there practice changes? How will the frontline caregivers beyond those attending the RCA team meeting be involved? The clinical content experts attending the RCA team meeting will help determine the selection of the latent root to address, and the recommendation most likely to eliminate the latent root cause. Additional team members (performance improvement, quality, clinical education, clinical informatics, among others) can help decide the best improvement strategy to achieve the recommendation.

The implementation plan should include verification and validation methods for any new or changed processes. That is, how will the team know the recommendation is implemented? And how will they know the recommended change is having the intended impact?

In some cases, the RCA team may be constrained by time and unable to develop an implementation plan. At a minimum, the RCA team needs to decide the recommendation. The team that has conducted the analysis of the unintended event using the Logic Tree is best positioned to recommend the specific action to take. Implementation, verification, and validation are sometimes completed in the RCA team meeting. However, if necessary, they may be developed in a subsequent meeting. A separate improvement team may be chartered. It is the responsibility of the Executive Sponsor of the RCA to assure the next steps are completed in a timely way.

5.4 RIGOR TESTING

The action plan should be rigor tested by the RCA team before approval. Harold developed a rigor-test checklist for RCA team use. The more these criteria are met, the stronger the action plan. (See Table 5.2 for the Action Plan Rigor Test.)

The RCA facilitator and Executive Sponsor collaborate to query the team before concluding the meeting. To assure that there is support, it can be helpful to ask each team member whether they agree the action plan is strong enough to eliminate the selected latent root cause. If the RCA facilitator, Process Owner, and Executive Sponsor are not convinced, the plan will eliminate the latent root, it is best to address the concerns at the meeting.

5.5 ACTION PLAN TEMPLATE

We developed an action plan template, which is a Word document that can be completed in real time with the RCA team. (See Table 5.3 for the Action Plan template.)

The template includes basic information: RCA date; due date for completion; names of the Process Owner, Executive Sponsor, and RCA facilitator. There is space for a brief description of the type of Event and Mode, usually one sentence. A recommendation is recorded. This should be limited to one or two action items. The top of the template also contains space for documenting metrics of the outcome, process, and balancing measures. The bottom of the template focuses on implementation: how the action will be rolled out; what are the processes, tools, roles, and communications needs; verification and validation; and the process for observation and knowing whether the action achieved the intended outcome. Columns are used to document the

---

TABLE 5.2
Action Plan Rigor Test

The more of these criteria that are met, the stronger the action plan becomes. This supports our goal of achieving "No Repeat Events."

1. Does the recommendation directly impact the latent root?
2. Is the recommendation "SMART" (Specific-Measurable-Actionable-Realistic-Timely)?
3. If the new process is dependent on a person/role or time(s)-of-day, is there a fail-safe backup to ensure it's done? (Must have to be viable 24/7.)
4. Does the recommendation survive changes in personnel?
5. Is the recommendation supported by data and/or best practices (literature/research)?
6. Does the recommendation meet the Shingo "Zero Quality Control" criteria?
   a. Source inspection – "checking your own work"
   b. 100% inspection – "checking all of the work, not a sample of the work"
   c. Immediate feedback – e.g., "I notice you didn't wash your hands"
   d. Mistake-proofed (Poke Yoke) – (e.g., medical gas fittings are different so gases cannot be mixed)

Additional Considerations
7. If a new process is recommended, ensure Operator Standard Work Instructions are written.
8. Assess the financial impacts by considering ROI (return on investment) of recommendation.

TABLE 5.3
Action Plan Template

Department/Unit:		Metrics	Current	Target	Target Met? Y/N
RCA date: Due date: Event Summary: Latent Root:	Executive Sponsor:	Outcome *What are we trying to achieve with this process? Is it having the effect we intended? Think specifically about the process, versus eliminating the event.*		(Zero/100%)	
Latent Root Category: Recommendation: *What specific action do you plan to eliminate the Latent Root Cause? Limit 1–2 action items.*	Process Owner: RCA Facilitator:	Process *What is the process of the action plan? Are we doing what we planned?*			
		Balancing *In implementing this action plan, are there unanticipated outcomes we can identify?*			

ACTION ITEM	OWNER	DUE DATE	STATUS	NOTES	SUSTAINABILITY (FOR EXECUTIVE GEMBA WALK POST-CLOSURE)		
					30	60	90
Implementation Plan				*Helpful tools:* Training Matrix Standard Work Template			
Verification and Validation *Verification – Did we implement the right thing, in the right way, to the right people? Tied to process metric.* *Validation – Did our action plan achieve our intended outcome? Tied to outcome metric.*				*Helpful tools:* Process Observation Form Process Observation Tracker Abnormality Tracker Gemba Rounding Calendar			
Closure Huddle							

owner, due date, status, notes, and dates for 30/60/90-day reviews. Also included is a row for documenting plans for conducting a Closure Huddle. The Closure Huddle is the gathering of those involved in the unintended event, from its initial identification through those involved in the RCA team meeting and action plans. It is an opportunity to close the loop and recognize the contributions of caregivers to improving safety.

The value of filling out the action plan template in real time with the RCA team is the ability to print and post it promptly in the department. There is value in making the action plan as visible as possible. It is intended to be a living document. That is, it should communicate the intended safety improvements beyond the immediate RCA team members. It can be used to tell the story of the RCA outcome. It can be referenced in huddles and department meetings. Progress on implementation can be tracked and shared. Setting a date for a Closure Huddle helps assure it takes place.

5.5.1 ACTION PLAN METRICS

Our approach to action plan metrics is consistent with the approach we advocate taking for action planning. There is correspondence between the metrics and the elements of the action plan: the recommendation, and implementation plan with verification and validation. We begin with the outcome measure. The outcome measure directly corresponds to the action recommendation. If the recommendation is achieved, what is the predicted result? This measure is either 100% or zero. That is, the outcome is intended to always prevent the latent root cause from reoccurring. If the RCA team is unable to complete an implementation plan at the time of the team meeting, they should at a minimum define the outcome measure.

A process measure (or measures) should be used to verify the implementation of the recommendation. Is the improvement strategy on track? Is the new or revised process documented? Is standardized work written? Are caregivers receiving the training required? Is needed communication taking place as planned? These are examples of process measures.

A balancing measure is used to monitor for possible unintended consequences of the action plan. Predicting a possible undesirable impact of the action plan, the balancing measure will gauge whether this is occurring. For example, an RCA team recommended adding an order for emergency physicians to write when certain patients were to be admitted. A balancing metric could be the number of clicks required, if too many, it is less likely the new order will be entered.

5.6 IMPROVEMENT TOOLS

Improvement methods include a whole host of tools and forms. We have found that using a few simple tools to support the RCA action plan is best. Following are those we have found helpful most often.

5.6.1 STANDARDIZED WORK

Implementing an RCA team recommendation very often means redefining how the work is done. A template for documenting standardized work helps support a new process or modifications to an existing process. At its most basic, standardized work should define the steps in the process, in the order they are performed, and how long each should take. It should be written with the minimum necessary detail, and it's helpful to use images and photos when possible. The standardized work steps can then taught using the TWI Job Breakdown method. Ideally, a caregiver unfamiliar with the work should be able to pick up the standardized work instructions and accomplish the task. New standardized work can be posted along with the RCA action plan on a unit huddle board for visibility and to help socialize the implementation. Examples of a Standardized Work template and a TWI Job Breakdown template are shown in Table 5.4.

TABLE 5.4
Standardized Work and TWI Job Breakdown Templates

Standardized Work		
Process:		
Step	**Description**	**Duration**
1		
2		
3		
4		
. . .		

Training within Industry Job Breakdown Sheet		
Important Steps	**Key Points**	**Reasons**
The major steps in the process that advance the work	Anything in an Important Step that might: 1. Make or break the work 2. Injure the worker 3. Make the work easier	The "Why" for the Key Points

5.6.2 TRAINING MATRIX

One of the challenges with RCA action plans is spreading the information to all those who need to know it. A training matrix is useful to track who has received the necessary orientation or training to the new standardized work. A posted list helps frontline caregivers and leaders track the status of implementation. See Table 5.5 for a Training Matrix template example based on the Benner/Dreyfus model.

5.6.3 PROCESS OBSERVATION

The purpose of process observations is to verify that caregivers are following the standardized work. The process observation form allows capturing information while observing the process being performed. It includes key steps in the standardized work. The form notes whether or not the steps are performed correctly and can guide the observer with opportunities to coach. It provides visibility to the progress of implementation and helps with trending of results. The RCA action plan implementation strategy may have indicated how many observations need to occur to be confident the new standardized work is fully implemented. Process observations show progress toward the new process adoption. See Table 5.6 for a Process Observation template.

5.6.4 ABNORMALITY TRACKER

When process observations note a deficiency in the performance of the standardized work, the abnormality tracker is a way to document the reasons for the deficiencies. It can show which steps are being missed and how frequently or steps that are not possible to perform correctly due to barriers. Each step in the standardized work can be listed across the bottom of the form. The date of each observed abnormality is added to the column. Alternately, each column may have a unique barrier or cause for the standardized work not being followed. The date of each observation of the

---

TABLE 5.5
Training Matrix Template

Training Matrix						
Unit/Department:						
Process:						
Skill Levels:	Novice, Advanced Beginner, Trained, Competent, Proficient, Expert					
	Task/Skill/Responsibility					
Caregivers						

TABLE 5.6
Process Observation Template

Process Observed:			Date:
Observer:			
Step	Key Step (see Standardized Work Instructions)	Done? Y/N or N/A	Comments
1			
2			
3			
4			
. . . .			

TABLE 5.7
Abnormality Tracker Template

Process Observed:			Unit/Department:		
Dates/Period:					
Dates of Occurrence					
Reasons for Deficiency					

abnormality is added to the appropriate column. Either of these forms for tracking abnormalities helps make visible where most of the problems occur, a Pareto Chart, and where the focus of refinement of the process is needed. See Table 5.7 for an Abnormality Tracker template example.

5.6.5 ROUNDING

Executive Sponsors, Process Owners, RCA facilitators, and others with content knowledge of the RCA action plan can support its implementation by rounding on a regular basis. This is purposeful rounding. Leaders can join unit huddles and inquire about the progress on implementing action plans. The visibility of the plan and tools provides data. The leaders can then focus their attention and help address barriers to implementation.

5.7 SUMMARY

Effective action planning is essential to prevent reoccurring unintended events. Incomplete or ineffective action plans are amongst the greatest risks to RCA. Focused actions to eliminate a single Latent Root cause and thereby preventing a reoccurring unintended event are required.

The action plan should be treated as a living document. Along with the tools used for implementation, the action plan should be visible and reviewed regularly. Process measures can be tracked using daily management systems such as unit huddles, and with frequent reviews, barriers to implementation are more easily identified and addressed.

5.8 QUESTIONS TO CONSIDER

1. What does a survey of the last year's RCA action plans in your organization reveal?
2. What barriers to fully implementing action plans do you observe?
3. How visible are action plans to organizational leaders, to know whether they are implemented and effective?

NOTES

1. National Patient Safety Foundation, *RCA2: Improving Root Cause Analyses and Actions to Prevent Harm* (Boston, MA: National Patient Safety Foundation, 2015).
2. "Oregon Patient Safety Commission Strength of Action Plans," www.oregonpatientsafety. org/news-information/adverse-event-investigation/building-a-strong-action-plan/494, accessed December 29, 2020. See also "The VA Root Cause Analysis Toolkit," 26–28, www.patientsafety.va.gov/docs/joe/rca_tools_2_15.pdf for Action Strength and a table of Action Hierarchy.
3. J. Reason, "Human Error: Models and Management," *British Medical Journal* 320 (2000): 768–70.

6 RCA Facilitation

The intent of this book is to equip those who facilitate healthcare RCAs with a deep understanding of the investigative tool. Our aim in this chapter is to pull together the building blocks necessary for facilitating a thorough and credible RCA. We will discuss the responsibilities of the RCA facilitator and of the Executive Sponsor. We share some practical tips on facilitation and ideas for skill development. The exercise at the end of the chapter will help cement the understanding of the Logic Tree methodology.

6.1 ETHICS OF INQUIRY

At the time of this writing, we are in the midst of the COVID-19 pandemic. The author and historian of an earlier pandemic, John M. Barry,[1] in *The Great Influenza*, described his approach to inquiry in an address to Vanderbilt University. Barry proposed the ethics of inquiry requires four questions to be asked in the pursuit of truth: *what happened, how, why*, and *so what?* "*What happened?*" necessitates decisions about what is included in the narrative and what is left out. Even if we are never fully able to know the truth, we are obligated to get as close to it as possible. How do we tell the story of what happened? "*How?*" is the mechanism by which the unintended events occurred. Scientific inquiry is the method for answering this question. Some of the heroes in his book showed the conviction and perseverance necessary for scientific proof. "*Why?*" seeks to understand the situation the persons confronted. What was the context? And finally, "*So what?*" seeks to understand how it matters to help set priorities for the future.

The pursuit of truth is the responsibility of the RCA facilitator. The Logic Tree helps us move from an understanding of the facts, what happened, to a deep inquiry into how it happened. Knowing the context for the decisions made along the way brings into light the underlying sources of error. RCA is the inquiry from the facts, to employing the scientific method of hypothesizing and seeking evidence, to re-prioritizing how work is done in the future. All is in the service of preventing harm.

6.2 PRACTICAL ASPECTS FOR FACILITATION

We have experienced many adaptations of our RCA facilitation as we have continued to learn and improve over the course of 15 years. In this section, we share observations and recommendations for several aspects.

6.2.1 PAPER BASED OR SOFTWARE

In Chapter 2, we shared our journey to adopting the Logic Tree methodology and using the PROACT software to support RCA. One clear lesson from the journey is that methodology is the key, whether the RCA is done with paper, commonly available software (typically Excel or Visio), or with software specifically designed for RCA (such as PROACT). We began our use of the Logic Tree using half-sheets of paper for the boxes and displaying them on a wall. It was easy to move the half-sheets around, and they were easily read. This requires wall space large enough to accommodate the Logic Tree. A few times when there has been a technology failure, teams have reverted to using paper to create a Logic Tree and were able to carry on with the RCA. This low-tech approach is easy. The downside is saving the Logic Tree usually requires transcribing it in some smaller form. This means additional steps for the RCA facilitator and possibly delays in the availability of the Logic Tree. The time it takes to complete the RCA can be impacted.

Software as the alternative to paper can eliminate re-work. Learning the software, though, does take practice. It can be challenging to facilitate an RCA team while still trying to gain competency in the software used. We have found that it pays off to spend time practicing in the software outside of team meetings. The exercises in this book and cases you know can be good material for practice sessions. It helps to have at a minimum the ability to create boxes, cut and paste, copy, and delete. Another lesson from experience is to learn how to "undelete," just in case.

Software that is designed to support RCA (such as PROACT) may allow storing the associated documents, information, and evidence from the investigation in one place. This can eliminate the need for multiple files. Some software programs also support action planning, tracking results, and communicating. A decision about whether to use RCA software goes beyond the ability to create and document the Logic Tree to include building a cohesive infrastructure to support RCA and patient safety.

6.2.2 Skill Development

There are multiple skills required to successfully facilitate an RCA team. At the top of the list of necessary skills is to understand the methodology of the Logic Tree, the use of deductive reasoning, and being able to accurately identify Physical, Human, and Latent roots. For those using software, there is also the skill required to run the software in real time while facilitating a team. Group facilitation skills are also essential.

Several approaches can be used to practice creating Logic Trees. An easy way to practice is with Post-It notes. Start with creating the Top Box using the exercises in this book or other cases you know of. Add hypotheses, and work your way down through the Physical, Human, and Latent roots. Watch for the correct sequence of roots and use the chain of causation to test your Logic Tree. It can be helpful when practicing to work with a partner. One person can work on building the Logic Tree based on information gleaned from interviewing a partner. This adds the dimension of using the question, "*How could?*" as you dig down. Include practicing the elimination of hypotheses based on the evidence.

Another type of practice can be for the learner to attend an RCA team meeting and work on a Logic Tree separately. This allows experiencing building the Logic Tree in real time as information from the team is being shared. We recommend letting the RCA team know at the beginning the intent of this activity, so it is understood and not a distraction.

It is our experience that RCA team facilitators may come out of the team meeting having captured critical data and identified latent roots, but without the chain of causation being clear. A helpful practice and way to increase skill is to spend time after the RCA team meeting reviewing the Logic Tree. Check to see whether the hypothesis statements were worded well. Are the Physical, Human, and Latent roots in the correct sequence? Trace the chain of causation up from the latent roots. Don't be surprised if there is a need to reword some of the boxes and move boxes around so the deductive reasoning is sound.

6.2.3 Shared Facilitation

We have found it important to have an Executive Sponsor and RCA facilitator team up in RCA team meetings. The sponsor helps set the tone of appreciative inquiry and a just culture and holds the team accountable for achieving a thorough and credible result. The role of the RCA facilitator can also be shared. In the early stages of leading RCAs with the Logic Tree methodology, some have found it helpful to divide the tasks of leading the discussion and documenting the Logic Tree. If using paper, one person can write and post the half-sheets while the other facilitates asking *how could that happen*? The same is true if using software. One leader can document as the partner leads the questioning.

Our aim has been to achieve a level of proficiency in using the software such that an individual facilitator can direct the inquiry and build the Logic Tree in real time. This eliminates delays caused by communication between the facilitators. The team can move ahead without interruption for the facilitators to catch up with one another. Achieving this takes time and experience, but in our experience is a best practice.

Another important element of shared facilitation is the need to monitor the RCA team's mood and engagement. It can be challenging for a leader who is guiding the inquiry and documenting the Logic Tree to read the body language of the group. Having a partner who is aware of team members who may not be contributing, or are dominating, and can help facilitate full participation is valuable. A second leader can help bring teams back who have gone down the proverbial rabbit hole. If the RCA team is impacted by high emotions, having two or three leaders prepared to express support and help guide the team forward in the analysis is useful. Regardless of how the roles are divided, it is a helpful practice standard to partner in facilitating RCA team meetings.

6.2.4 Team Focus

Keeping an RCA team on task is important. Some key tactics include the agenda, ground rules, and the physical setup of the conference room. The agenda for an RCA team meeting is quite simple. It includes introductions (if needed), a welcome and statement on the importance of the work and goal of achieving improved patient safety, setting team ground rules including confidentiality, reviewing a chronology of the facts, conducting an analysis of the case with the Logic Tree, and developing an action plan. Sometimes a just-in-time training is provided before the team's work begins. A quick review of the agenda helps orient the team to the task and allows easier redirection back to the stage of the agenda being worked if needed. A trap healthcare RCA teams easily fall into is jumping to problem-solving without having completed the analysis. Reminding the team of the importance of confirming the system level problems before moving into action planning helps retain their focus. Another tactic sometimes used is to have post-it notes available to the team. Let them know if they have an idea or recommendation while the chronology is under review or the Logic Tree is being developed that they can jot it on the note. When the time comes for action planning, the ideas may be shared. Another tactic is to use a "parking lot" or "additional recommendations" page. It is not uncommon for teams to generate multiple ideas, some of which will not directly mitigate the risk of repeating the unintended event. To respect the ideas and capture them for consideration later, they can be documented on a list for the Process Owner. They may be excellent ideas for process improvement and taken up later.

The team can be invited to generate a list of ground rules. Or, for expedience, a standard set of ground rules can be adopted for all RCA team meetings. A good reason for covering ground rules is the conversation they can generate. Every team needs to consider in some way: the value of encouraging team members to speak up, being respectful in communication, being committed to improving systems over blaming individuals, and how confidentiality will be handled.

6.2.5 ACTIVE FACILITATION

It is critical that the RCA facilitator actively guide the RCA team. Thorough and credible RCA results are predicated on a standard approach that insures the rigor of the process. This is true throughout the agenda. The tone established at the beginning of the RCA team meeting will help ensure the information needed to understand how an unintended event occurred will be shared. Communicate an attitude of respect, openness, and assuring the psychological safety of the team. This may be critically important for team members who feel responsible for having harmed a patient. One way to address the fears team members may harbor is to talk about the purpose for the RCA. Share that the goals are to understand how the unintended event occurred, find ways to prevent it from happening again, and share the lessons learned. Our assumption is that involved caregivers are caring and competent people who did not intend to harm a patient. We assume the sequence of events that lined up resulting in their experiencing an unintended event could also happen to other equally caring and competent caregivers if the systems failures are not addressed. The team's contribution to this effort is needed and valued.

The unintended event chronology was carefully developed with consideration for what facts need to be covered, often following the flow of the patient. When reviewing the chronology, it is advisable for the RCA facilitator, not the team members, to talk through the sequence of events. We have found that inviting team members to "share their story" risks moving into explanations and causation. At this step in the process, we just want to get a common set of facts on the table. The RCA facilitator, or another designated team member, should review the chronology. Once this is done, the team can be invited to correct any misinformation, add any important points missed, and identify any other sources of data that may not have been consulted when preparing the chronology. This approach honors input from team members while avoiding possible pitfalls.

Whether using paper-based or software-supported versions of the Logic Tree, an advantage of either is the focus it brings to the team. It is important the conference room setup allows the team to all see and focus on the display of the Logic Tree. We have found that having the facilitator for the Logic Tree be positioned near the front alongside the display helps maintain team focus. The team can both follow the leader and see the Logic Tree. A second facilitator might be across the conference room in a position allowing observation of the whole team. A potential pitfall for the RCA facilitator is to give up actively guiding the team and become a note taker for a conversation happening between team members. While this may be valuable conversation, it lacks the discipline and rigor of analysis. It is vulnerable to opinion expressed by team members higher on the authority gradient, or the most vocal. As facilitators, be aware of moments in which the team stops focusing on the display of the Logic Tree and turns toward one another – verbally guide them back to the point on the Logic Tree under discussion.

The novel approach to action planning we have developed is foreign to many healthcare teams. RCA action plans have often been seen as opportunities to escalate a priority of pet projects or previously delayed recommendations. Action Plans risk becoming so heavy with recommendations, even when well intentioned, that they are doomed for the failures previously described. The RCA facilitator must help the team focus on the critical question of which latent root cause will be the focus of action, and what single recommendation is aimed at 100% implementation reliability. This requires guidance and discipline for the team. As previously suggested, additional recommendations can be captured and given to the Process Owner for future consideration. However, for the purposes of the RCA action plan, the team should be focused on active facilitation to increase the likelihood of success in preventing reoccurring unintended events.

6.2.6 FACILITATING THE LOGIC TREE

Focused attention by the team on the Logic Tree is important. We have learned some lessons on how to go about leading the team to create the Logic Tree. The starting point is the Top Box. We have recommended creating the Top Box in advance and displaying it when the time comes for the analysis. The decision about the Event type and Mode (or Modes) should be factually based and not requiring discussion. A beginning point is to acknowledge the reason the team is gathered is the serious safety event, close call, or other type of Event that led to chartering the RCA. The Mode is a factual statement and should be consistent with the chronology of the facts just completed. Sharing the Top Box helps guide the team.

We previously discussed whether to pre-populate the Logic Tree with Physical Root hypotheses or to list nothing as the initial hypotheses. In either case, the question to the team is, "*How could this (Mode) have happened?*" The team can be asked to confirm the hypotheses displayed from the initial investigation or to generate a new list. Always leave open the possibility of adding additional hypotheses at this level. Having a blank box visually cues the team into the possibility. We have found it best to guide the team in generating this initial list of hypotheses by asking, *"How else could this have happened?"* before moving down the chain.

When the initial hypotheses list has been completed, it helps to guide the team in an orderly process for investigating each hypothesis. For convenience, we start on the left side of the hypotheses list and ask *how could this (hypothesis) have happened? What does the evidence show?* If there is evidence to confirm the hypothesis, then the next level of hypotheses is generated. *How could that happen?* With practice, you will see when these top hypotheses are confirmed as true, they are usually Physical Roots. Keep drilling down that root system of the Logic Tree until you reach a decision point. Continue to generate hypotheses on how the decision in error was made. *Why was that decision made?* Below the human root level, you will begin to identify Latent Roots. Sometimes there is more than one Latent Root that contributed to the Human Root. Sometimes a Latent Root caused another system-level vulnerability. In other words, a Latent Root leading to another Latent Root, which led to a Human Root, and ultimately a Physical Root. In our passenger train derailment exercise, for example, recall the evidence that budget cuts, a Latent Root, led to changes in the preventive maintenance schedule, a subsequent Latent Root.

In cases of Latent Roots leading to Latent Roots, teams may wonder when to stop documenting Latent Root Causes. Practically speaking, we suggest going as far in documenting Latent Roots as those within the span of control of your organization. Think, for example, of Latent Roots related to drug manufacturing. Changes in products (a Latent Root) may lead to vulnerabilities in your organization's procurement, storage, labeling, or communication with end-user systems (also Latent Roots). The RCA will benefit from rigorously identifying Latent Roots with direct organizational control. Your organization may be able to communicate the risks and hazards created by external manufacturing changes (such as through FDA, Safe Medical Device reports), but is usually incapable through RCA of creating countermeasures to the external causes of those changes.

A eureka moment for the team can be when you have documented a causal chain down to the Latent Root level. It is confirming to the team for the facilitator to reflect back to them this chain of causation by verbally walking up from the Latent Root to the Mode. This demonstration of the deductive reasoning process clues the team into the methodology and guides them for the remainder of the analysis, typically accelerating the remainder of the Logic Tree development.

6.3 SUMMARY

Part I of the book has aimed to build the capacity to lead an RCA team in conducting a rigorous and credible RCA, and to develop an effective action plan to mitigate future harm. As leaders of this powerful tool for investigation, we are in a strong position to understand the various forms of analysis and how RCA fits among them. We can facilitate early investigation and the gathering of information needed to evidence the findings of the RCA. The skills of chronology development, of framing a Logic Tree Top Box, and of guiding teams in hypothesizing and asking what the evidence shows, lead to the answer for how a healthcare unintended event occurred. Guiding teams to identify Latent Root causes and not stopping at the Human Root level is necessary to address the vulnerabilities in our systems of care delivery. Helping caregivers choose the optimal latent root to eliminate through an action plan, and the ability to measure the implementation and outcome of the recommendation leads to better outcomes. Executive sponsors and RCA facilitators have a moral responsibility to assure the thorough and credible analysis and action planning to prevent reoccurring unintended events. Doing so is our compassionate response to the harm befallen on our patients and their loved ones, as well as our caregiver colleagues.

In Part II, we transition from describing the knowledge and skills needed to conduct an RCA to the ways in which we can champion the effective use of RCA in our healthcare organizations.

6.4 QUESTIONS TO CONSIDER

1. Will you take a team approach to facilitating RCAs? With whom can you partner to practice the Logic Tree methodology?
2. Will you utilize a software or paper-based approach? What steps will be required for implementation?
3. What organizational supports will be needed to implement thorough and credible RCA?

6.5 EXERCISE 5: LOGIC TREE 2

This exercise is a continuation of the exercise in Chapter 3. Your team has been assigned to investigate and conduct an RCA after the cancellation of a surgery due to the lack of a necessary supply. A special mesh, preferred by the surgeon for certain difficult abdominal repairs, was not available in the operating room. Unfortunately, the patient had already been sedated. The case was canceled and rescheduled.

DIRECTIONS

Additional information has been obtained. Using the following facts, create a Logic Tree.

- Label the Event: "Logic Tree Training Exercise #2."
- Label a Mode based on the condition that to successfully start the surgery, the special mesh should have been in the operating room *and* supplies should have been confirmed before anesthesia was induced
- Review the results of the interviews and information gathered by your team. Label each item as a Physical root, Human root, or Latent root
- Construct a Logic Tree

Facts:

- Ms. Fortune initially saw her surgeon, Dr. Dogood, in the clinic. Dr. Dogood recommended surgery to revise an abdominal repair that was leaking and discussed possible risks and benefits of the procedure. Ms. Fortune agreed and the procedure was scheduled for the next week.

- Dr. Dogood asked her staff to complete the consent form and to schedule the surgery. The office nurse filled out the consent form for Dr. Dogood to have Ms. Fortune sign.
- The office clerk faxed a copy of the request for surgery time, along with the consent form, to the hospital scheduler.
- When reviewed, the surgery schedule does not indicate any special equipment or supplies are needed.
- The forms received from the surgeon's office were not kept. However, the scheduler tells you that if they included any information about special mesh, this would be noted on the hospital surgery schedule.
- The scheduler adds Ms. Fortune to the surgery schedule.
- The surgery charge nurse reviews cases for the next day, including Ms. Fortune's.
- The surgery charge nurse tells you there are occasionally errors on the surgery schedule when information is missing or wrong. She doesn't know what happened in this case.
- The surgery tech tells you that when it was discovered the special mesh was not on the back table, he called to the supply room. He was told they had one special mesh but that it was outdated and could not be used.
- The surgeon states the special mesh is only used in complicated re-operations. These cases are rare.
- The sterile processing manager tells you the special mesh is only used once or twice a year. Her records show two were purchased last year, and one was used.
- Sterile processing does not have a process for checking outdates on specialty supplies. The practice has been to re-order when the supply is used.
- The day of surgery, the circulating nurse brings Ms. Fortune into the Operating Room and confirms her name and procedure.
- Ms. Fortune is sedated by the anesthesiologist.
- The surgeon arrives at the room and leads a final time out. She asks whether the special mesh is available.
- The team realizes the supply is not available and contacts Sterile Processing.
- The case is canceled and rescheduled.

NOTE

1. J. M. Barry, "Hurricanes, Viruses, Politics and the Nature of Inquiry," Vanderbilt University address delivered in November 2006, accessed on YouTube April 6, 2020.

Part II

Root Cause Analysis Champions

We have done a deep dive in Part I into the facilitation of an RCA, from the identification of an unintended event and decision, to conducting an RCA through the implementation of an action plan. This is an important skill and core competency for patient safety professionals and others committed to eliminating patient harm. In Part II, we build on this foundation with additional knowledge to make RCA part of an integrated healthcare safety system. We outline standard expectations for RCA team participation to better enable the coaching of team members. The ways in which RCAs can fail are examined and countermeasures identified. How trends from RCAs can be identified and capitalized upon are shared. And how RCA fits into a broader array of improvement strategies for preventing harm is suggested. Finally, we share a curriculum for teaching RCA in healthcare. Knowledge of each of these building blocks, added to the foundation of facilitating effective RCAs, enables serving as a champion for RCA within healthcare organizations.

7 RCA Standardized Work by Role

Many caregivers have roles in rigorous and thorough RCA. Understanding and cultivating the contribution of key roles is a function of the RCA champion. In this chapter, we share standardized work recommendations for a number of these roles. We also suggest tactics for orienting team members to the standardized work.

What do we mean by standardized work? Standardized work (sometimes called Operator Standard Work Instructions) defines a set of tasks to be performed by role. It lists the tasks in the order in which they should be performed. The practice of defining standardized work developed from the Training within Industry approach that emerged in World War II to quickly teach women (think Rosie the Riveter) the processes needed for industrial manufacturing in support of the war effort. A set of instructions to complete a task, in the proper order, and written simply enough that a caregiver new to RCA understands the expectations of their involvement is the goal. Table 5.5 shows examples of Standardized Work templates.

7.1 STANDARDIZED WORK BY ROLE

Following are standardized work recommendations for several roles. The standardized work instructions should be adapted to the processes for RCA defined by each healthcare organization. We recommend including the following critical elements:

7.1.1 EXECUTIVE SPONSOR

The Executive Sponsor is a member of the Executive Team with accountability for the unit or department involved in the unintended event. The Executive Sponsor is responsible for supporting the RCA process and removing barriers as needed. Their standardized work is:

1. Receive notification of the unintended event.
2. Participate in prioritizing and determining if an RCA is indicated.
3. Conduct purposeful rounds in the involved department:
   a. Check in with manager, supervisor, caregivers to assure caregiver support is provided when needed.
   b. Confirm immediate countermeasures for safety are in place pending an RCA action plan.
4. Monitor progress and support early investigation as needed.
   a. Confirm team roster with the RCA facilitator.
   b. Review chronology in advance of RCA team meeting.
   c. Support Medical Staff involvement in RCA team meeting as needed.

5. Address barriers to the timely RCA team meeting and action planning.
6. Attend and participate in the RCA team meeting:
   a. Provide welcome and opening remarks.
   b. Confirm identification of Latent Root Causes.
   c. Use Rigor Test for approval of the action plan.
7. Monitor and support implementation of the action plan.
8. Conduct a Closure Huddle.
9. Communicate RCA results.
10. Monitor 30/60/90/180-day results.

7.1.2 PROCESS OWNER

The Process Owner is typically an operational leader such as a manager or supervisor with direct accountability for the processes involved in the unintended event. Their standardized work is:

1. Receive notification of the unintended event.
2. Participate in prioritizing and determining if an RCA is indicated.
3. Conduct purposeful rounds in the involved department:
   a. Check in with caregivers to assure caregiver support is provided when needed.
   b. Confirm immediate countermeasures for safety are in place pending an RCA action plan.
4. Collaborate on early investigation.
5. Remove barriers to caregiver attendance at the RCA team meeting.
6. Attend and support the RCA team meeting.
7. Collaborate on implementation of the RCA action plan.
   a. Post RCA action plan.
   b. Monitor and verify implementation using process metrics.
   c. Monitor and validate results using outcome metrics.
   d. Utilize improvement tools such as abnormality tracker, process observations tracker, and training matrix as needed.
   e. Report and escalate barriers to Executive Sponsor.
   f. Conduct purposeful rounds and coach new processes to stability.
8. Attend and participate in Closure Huddle.
9. Support 30/60/90/180-day results.

7.1.3 DIRECT CAREGIVER

Caregivers directly involved in an unintended event provide critical information to the RCA team meeting. Advance support may be needed to assure they are emotionally able to contribute to the RCA team meeting. Their standardized work is:

1. Participate in interviews and information gathered in advance.
2. Attend and participate in the RCA team meeting:
   a. Work with immediate leader to remove barriers to attendance at the RCA team meeting.

    b. Contribute to the chronology.

    c. Help identify Latent Root Causes.

    d. Participate in developing an action plan to prevent reoccurrence of the unintended event.

3. Attend the Closure Huddle.

4. Help communicate RCA results.

7.1.4 Physician Leader

RCAs involving providers benefit from the participation of a physician leader: Medical Director, Division Chief, or Medical Staff Officer. If needed, physician leaders can serve as a liaison to the physicians involved in the event. They can also provide information regarding medical decision-making, procedures, and best practices. Their standardized work is:

1. Receive notification of the unintended event.

2. Support the RCA early investigation as needed.

3. Interview involved provider(s) unable to attend the RCA team meeting and represent the information received.

4. Attend and participate in the RCA team meeting.

    a. Contribute to the chronology as needed.

    b. Help identify Latent Root Causes.

    c. Participate in developing an action plan to prevent reoccurrence of the unintended event.

5. Attend the Closure Huddle.

6. Communicate results of the RCA to Medical Staff as needed.

7.1.5 Subject Matter Experts

A few RCA team members may serve as Subject Matter Experts. Some such as operational leaders and clinical educators may provide clinical expertise and knowledge of best practices. Others such as informaticists, pharmacists, biomedical engineers, and facility engineers may provide specific content knowledge. Their standardized work is:

1. Receive notification of the unintended event.

2. Participate as needed in prioritizing and determining if an RCA is indicated.

3. Support as needed the RCA early investigation.

4. Attend and participate in the RCA team meeting.

5. Support implementation of the RCA action plan as needed.

6. Attend the Closure Huddle.

7.1.6 Support Services

A few RCA team members may serve in support of the RCA process and implementation of the action plan. These may include Quality leaders, Risk Managers,

Improvement leaders, Patient Safety leaders, and Patient/Family Advisors. In some cases, there may be parallel investigations being conducted, and collaboration among the disciplines is needed. Their standardized work is:

1. Receive notification of the unintended event.
2. Participate as needed in prioritizing and determining if an RCA is indicated.
3. Collaborate on RCA early investigation and coordinating parallel investigations as appropriate.
4. Attend and participate in the RCA team meeting.
5. Support implementation of the RCA action plan as needed.
6. Attend the Closure Huddle.
7. (Patient Safety leaders) track 30/60/90/180-day results.

7.2 IMPLEMENTING STANDARDIZED WORK

Training, orienting, and coaching caregivers to the RCA team standardized work by role is an important way to champion effective RCA. Here are suggested approaches you might take as the RCA Champion.

Training for effective RCA participation should ideally take place before needing to respond to an unintended event. Just-in-time training may also occur but is less likely to be comprehensive. An orientation to patient safety and RCA can be part of new leader onboarding. This is likely to be an overview of the safety program but can include a description of the approach taken in RCA. If the leader may later serve as an Executive Sponsor or Process Owner, standardized work instructions can be shared. Similarly, a general orientation to patient safety, event reporting, and the use of RCA to reduce harm can be included in an orientation for all new caregivers.

We have found a more effective form of training is to provide sessions dedicated to RCA and the roles of Executive Sponsor, Process Owner, and Subject Matter Expert. The scope of training may span from unintended event identification through action plan completion, or individual elements of the RCA process. The agenda should include a reflection on the commitment to reducing harm and why RCA is important. Provide an overview of the organization's philosophical approach to patient safety and RCA, noting the steps in the process: the interview process, the use of chronologies, the Logic Tree, and expectations for action plans. Introduce the importance of standardized work and invite participants to review their standardized work by role. An orientation to the Logic Tree is important. We particularly want Executives to understand the methodology, have faith in the process, and be able to support it with RCA teams. They too are RCA champions. It is valuable to orient leaders to action planning. Engaging them with why it is important to identify the one Latent Root Cause for elimination, and the value of limiting action plans to one or two recommendations will enable them to help support the process. Leaders knowing how RCA teams can fail and the countermeasures (described in Chapter 8) will enable them to help avoid failed RCAs.

Case studies are great for RCA training. Once the content listed for role training has been covered, breaking into small groups, and using cases and role-plays allows for interactive, shared learning. Adult learners will benefit from practicing the interactive skills, such as active listening, humble inquiry, identifying the Latent Root Causes, determining realistic metrics, and conducting Closure Huddles. If in-person training is not possible, consider interactive electronic training by video conferencing. More options for videos, podcasts, and other engaging forms of pedagogy are emerging. We've found individually viewed slide presentations to be least effective.

7.3 SUMMARY

Thorough and credible RCA involves a team. There are important, distinct roles that contribute to the overall success of the analysis. An important role for RCA champions is to support the understanding of these team roles through training, communication, and organization. Advance training will help ensure the ability to respond quickly and effectively to an unintended event, with greater likelihood of preventing the harm from reoccurring.

7.4 QUESTIONS TO CONSIDER

1. Which roles are currently involved in your organization's RCA?
2. If executives are not currently involved in RCA, what steps are needed to engage their participation?

8 Barriers to RCA and Their Countermeasures

Many healthcare organizations have utilized some form of root cause analysis for years, and yet continue to experience significant rates of patient harm. How can that be? Two possibilities exist. One is that the RCAs performed have not been successful. This possibility, considering the barriers to successful RCA and countermeasures to increase the likelihood of success, is the focus of this chapter. The second possibility is that RCA as a tool alone is insufficient to achieve the goal of zero harm. We explore that subject in Chapter 9.

RCA facilitators, Executive Sponsors, Process Owners, and all of healthcare leadership have a vested interest in successful RCA. How does RCA sometimes fail? To answer this, we will do an RCA of failed RCAs. Let us analyze *how could RCAs fail*?

8.1 HOW COULD RCAS FAIL?

For the Top Box of this Logic Tree, we will designate Failed RCAs as the Event, as evidenced by repeat events, the Mode. Let's explore three main hypotheses for how an RCA could fail: *Latent Roots were not found*; *Action Plans were not implemented*; and/or *Action Plans were implemented but not effective*. (See Figure 8.1 showing the Top Box and Hypotheses.)

Let's drill down on the first hypothesis, that the RCA did not identify Latent Roots. How could this happen? Four possibilities exist: the RCA suffered from an *informational problem*, an *organizational problem*, it *stopped at the Human Root* level, or was hindered by *bias*. Any of these problems, or a combination of them, may have resulted in the RCA not getting to the Latent Root causes. If this were to occur, any action plan could miss preventing a reoccurrence of the unintended event.

Continuing down the Logic Tree, how could there be an informational problem? Two hypotheses are: the RCA relied on opinion, and/or the RCA lacked verification of roots. How could there be an organizational problem? There could be barriers related to the timely scheduling of the RCA and the impact on memories. There could also be problems with the physical environment for the RCA team meeting, inhibiting the ability of the team. There could also be problems stemming from the culture of the organization and how RCA is viewed. How could the RCA stop at the Human Root level? This could be caused by poor facilitation stemming from a failure to follow the Logic Tree methodology. How could there be bias which prevents identifying Latent Roots? This could be caused by an incomplete RCA team, lack of RCA training, poor facilitation, and prior experience with similar errors.

If we follow the chain of causation for each of these, we can see the potential result of failing to adequately identify Latent Root Causes and, therefore, being vulnerable to reoccurring unintended events. Relying on opinion and the lack of

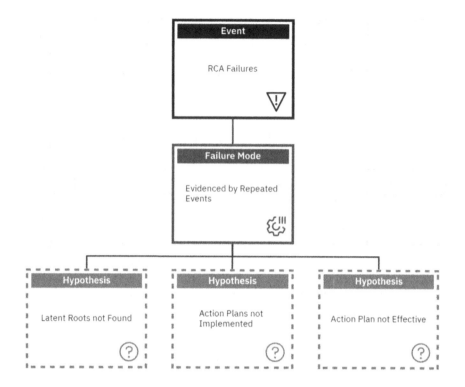

FIGURE 8.1   RCA Failure Hypotheses

Source: EasyRCA image used by permission of Reliability Center, Inc.

verifications causes informational problems. Delays in scheduling RCA team meetings, lack of adequate physical setup for the team to participate fully, and a culture more conducive to blame than identifying systems failures all have the potential to result in an RCA suffering from organizational problems to the degree that the Latent Roots are not identified. The failure of the RCA facilitator and Executive Sponsor to guide a team through the decision errors of the Human Root level to the Latent Roots will result in RCA failure. And, failing to address the many forms of bias that can influence a team may result in the failure to consider hypotheses that turn out to be factual. (See Figure 8.2 for the Logic Tree display.)

Perhaps an RCA team identified Latent Roots but failed to adequately address them with an action plan. Our second main hypothesis is that the RCA failure was caused by the failure to implement action plans. How could that happen? Three hypotheses are: the action plans were not implemented because barriers were not removed, the lack of an accountability structure, or constraints on the resources needed for implementation. (See Figure 8.3 for the Logic Tree display.)

Thinking back to the survey of RCA results conducted by David, each of these hypotheses was evident in the failed action plans.

The third possibility is the RCA failed to prevent reoccurrence of unintended events because the implemented action plan was not effective. That is, the plan

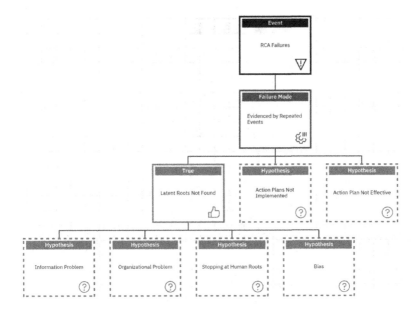

FIGURE 8.2 RCA Failure: Latent Roots Not Found

Source: EasyRCA image used by permission of Reliability Center, Inc.

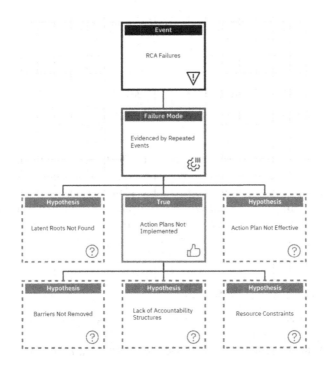

FIGURE 8.3 RCA Failure: Action Plans Not Implemented

Source: EasyRCA image used by permission of Reliability Center, Inc.

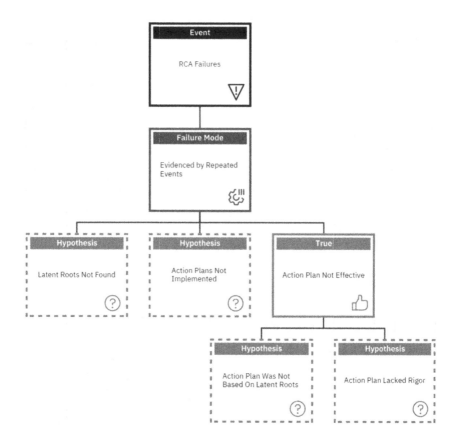

FIGURE 8.4   RCA Failure: Action Plan Not Effective

Source: EasyRCA image used by permission of Reliability Center, Inc.

was rolled out but did not have the intended effect. How could that happen? Two hypotheses are: the action plan was not based on a Latent Root, or the action plan lacked sufficient rigor. A fully implemented plan that doesn't prevent the Latent Root from repeating will not prevent future harm. An action plan that is not reliably implemented will continue to allow vulnerabilities in the Latent Roots and will not prevent reoccurrence of the unintended event. (See Figure 8.4 for the Logic Tree display.)

8.2 COUNTERMEASURES

Let us explore potential countermeasures for each of these potential failures. Table 8.1 suggests countermeasures.

Part I of this book detailed each of these countermeasures. Understanding how RCAs can fail allows implementing countermeasures for prevention. One role for RCA champions is to help facilitate the understanding of these possible sources of

TABLE 8.1
RCA Failure: Table of Countermeasures

RCA Failure	Root Causes	Countermeasure
Latent Roots Not Identified	Information Problem ▪ Relying on opinion ▪ Lack of verifications	RCA facilitation ▪ Use of verifications ▪ Adequate time
	Organizational Problem	Block schedules/rooms for RCA Ensure working equipment Cultural support from leadership
	Stopping at Human Roots	Facilitation of the Logic Tree methodology Cause-and-Effect Test
	Bias	Team composition Preparation/team RCA training Objective and broad hypotheses generation
Action Plan Not Implemented	Barriers not removed	Executive sponsor engagement
	Lack of accountability structures	Create visibility Track measures
	Resource constraints	Prioritization by executive sponsor
Action Plan Implemented But Not Effective	Action plan was not based on Latent Root	Use Cause and Effect Test before approval of action plan Use Rigor Test before approval of action plan Perform the "Check" part of PDCA
	Action plan lacks rigor	Use Rigor Test before approval of action plan Perform the "Check" part of PDCA

RCA failure and to advocate for the countermeasures as part of the overall safety management system.

Let us review the RCA failure countermeasures.

8.2.1 RCA Facilitation

Understanding the RCA methodology and skill in its use will enable RCA facilitators to guide RCA teams to rely less on opinion and more on verifiable information. This requires the ability to guide the team in creating the Logic Tree, and sufficient time and access to the information needed for verifications. These practices will move the analysis beyond the Human Root level to the Latent Root causes. This may be a shift in the organizational culture requiring both advocacy for the systems approach to preventing harm and the ability to work compassionately with impacted caregivers. Using the cause-and-effect test, walking up the chain of causation with a team, will help ensure the RCA has successfully reached the Latent Root causes of the unintended event. Facilitating the action plan Rigor Test ensures agreement that the plan addresses Latent Roots and will achieve permanent elimination of the Latent Root.

8.2.2 EXECUTIVE SPONSORS

Executive Sponsors play a critical role in preventing RCA failures. Their attention throughout the RCA process is necessary. Providing training for Executive Sponsors on their standardized work and its importance can help set the stage for success. Executive Sponsors can be great partners in understanding the use of the Logic Tree and creating an environment for teams to be fully engaged and open to drive the analysis to the Latent Root level. They should approve action plans and assure the resources needed for implementation are available. If barriers come up, they take the lead in addressing them.

Executive Sponsors should work with Process Owners to make the results of an RCA visible. Share the story within the organization, post the action plan, and charter an improvement team if needed to develop the new standardized work or process. Executive Sponsors are instrumental in monitoring the implementation of the RCA action plan, and its verification and validation. They need to be courageous enough to resist closing an action plan if it is not evidencing effectiveness. The Executive Sponsor is the person who can pull an RCA team back together to re-analyze the root causes or improve the action plan.

8.2.3 ORGANIZATION

Several countermeasures are based on the agreements on the processes that support RCA. These include the practice of scheduling and holding block RCA times and conference rooms. It means having the equipment necessary for the RCA team meeting. It also means prioritizing participation in RCA at all levels, including freeing up frontline caregivers to attend RCA team meetings, as well as executives.

The importance the organization places on safety and preventing harm, from the board on down, will be manifest in the RCA experience. Establishing a culture of psychological safety and justly evaluating caregivers sets a tone, a culture of safety, in support of effective RCA.

8.2.4 RCA EVALUATION TOOL

How can we evaluate the effectiveness of our RCAs? The most important measure is whether we have reoccurring unintended events – the outcome measure. In addition, we can establish monitoring systems, process measures, for action plans at the 30/60/90/180-day marks. This will help bring visibility to cases where the action plan is not completed or does not achieve the intended outcome.

For many years, we lacked an RCA evaluation checklist that would identify shortcomings and opportunities for improvement. From our experience, we have developed one to help identify strengths and improvement opportunities for the preparation and facilitation of the RCA team meeting, the Logic Tree analysis, and the action planning phase. (See Table 8.2 for a template.)

The checklist can be utilized in several ways. It can be used for self-evaluation after completing the RCA team meeting, to help guide critical self-reflection and opportunities for growth. Another approach is for the RCA Facilitator and Executive Sponsor to review the checklist together, which can help build collaboration. For the RCA champion, the checklist can be used to support others who are facilitating RCA team meetings. Using the checklist can bring a measure of objectivity to RCA feedback. The goal is continued improvement, and to guide discussion on the next step in learning about conducting effective RCAs. Teams of RCA facilitators might use the checklist with peers. Invite a colleague to attend an RCA team meeting to observe and provide feedback. This has the advantage of not only offering feedback but also increasing the depth of understanding by the peer reviewer in thinking through the evaluation.

TABLE 8.2
RCA Evaluation Template

RCA Case:	Date:	
RCA Facilitator:	Evaluator:	
Elements	Strengths	Opportunities
RCA preparation and facilitation:		

Room set up appropriately.
- Equipment set up effectively, including video or phone, as indicated.

Appropriate attendance achieved.
- Including leaders and support members.

Frontline caregivers present and encouraged to actively participate.
- Caregivers invited to share and ensure everyone has a chance to share.

Agenda reviewed.
- Clear welcome, introductions, ground rules, and process overview provided.

Disciplined data collection performed.
- Timeline prepared, interviews conducted, policy/ procedure reviewed, and Subject Matter Expert utilized, as indicated.

Strong facilitation skills exemplified.
- Conflict management, diffusing tension, mitigating behavior concerns, encouraging open and safe environment, humble inquiry, and engaging active participation.

(Continued)

TABLE 8.2 (Continued)

RCA Case:		Date:	
RCA Facilitator:		Evaluator:	
Elements		Strengths	Opportunities
RCA preparation and facilitation:			

Logic Tree Analysis

The event/mode well described and accurately identified.
- Confirmed with team.

The hypotheses were outlined as provable or disprovable statements.
- Team invited to explore hypotheses not populated on tree.

Disciplined data preservation performed.
- Providing visual aids, physical equipment, and pictures available, as needed.

Physical, human, and latent roots were reviewed with the team during analysis.
- Roots are captured (not necessarily labeled) within tree and pathway to failure is evident.

Latent root causes clearly identified.
- Focused on underlying system and process issues.

Latent roots delineated from contributing factors.
- Pathway to failure articulated to delineate, as needed.

Latent root(s) with the most potential for preventing reoccurrence selected for action planning.
- Team guided to select 1–2 latent roots.

Action Planning

Team brain stormed opportunities for improvement.
- Clear transition from logic tree investigation to solutions brainstorming.

Rigor test elements met for action plan.
- SMART, eliminates latent root, greatest potential for success, based on best practice.

Metrics are identified.
- Accurate, achievable, preferably measurable.

Support of action plan confirmed with team.
- Executive Sponsor confirms.

8.3 SUMMARY

The history of reoccurring unintended events shows that, as an industry, our past RCA efforts have not fully achieved the goal of preventing harm. An analysis of how RCAs can fail predictably shows that Latent Roots in our processes not mediated will result in further failures. RCA champions can lead the way in mitigating these risks.

Partnering with executives and leaders across the healthcare organization will lead to better analyses, better action plans, and more effective RCAs.

8.4 QUESTIONS TO CONSIDER

1. How does your organization make visible the outcomes of RCAs?
2. What countermeasures to the barriers for successful RCA will you implement?

9 Strategies for No Repeat Events

Our goal is the elimination of reoccurring unintended events. Achieving 100% perfect care and zero harm means unintended events are rigorously reviewed to identify Latent Root causes and action plans are implemented to prevent them from reoccurring. Reliability in discovering the causes and preventing them from reoccurring is required for "perfect care." Highly reliable processes will prevent harm. In this chapter, we explore strategies that the organization and the RCA champion can use for preventing Repeat Events and progressing toward the zero harm goal.

9.1 REPEAT EVENTS DEFINED

What do we mean by "Repeat Events?" It is important we define terms. Our nomenclature for reoccurring unintended events is based on the levels within the Logic Tree: Event, Mode, and Latent Root.

Unintended events are classified following the healthcare organization's policy definitions. We have used Serious Safety Event, Close Call, Operational, and Quality/Improvement Opportunity. This level conveys *why* an RCA is being conducted. As noted previously, we trend at the Event level to measure progress toward zero harm.

The *Mode* is a specific description of fact (see Chapter 3). It is *what* is being analyzed. Linked to the specific fact description of the Mode is the category or type of case, again based on the healthcare organization's definitions, such as the NQF list of 29 "Serious Reportable Events" or the Joint Commissions list of "Sentinel Events." Likewise, the Latent Root level conveys *why* the unintended event occurred. These are categorized according to the healthcare organization's policy and procedures, which may be based on the AHRQ Common Formats, a Joint Commission list of commonly reported causes, or other sources.

We propose a definition of *Repeat Events* as *a reoccurrence of harm with the same specific facts of the Mode and the same Latent Root causes.* We will say more about this later, but first need to ask, how do we capitalize on the lessons of an RCA following an unintended event? The answer lies in how we utilize trending and how we spread action plans and in understanding how RCA fits into a larger framework of a safety-improvement system.

To spell this out further, the definition refers only to Serious Safety Events. While patterns of repeated Close Calls are important opportunities, for purposes of measuring improvement and progress against the target of zero harm, the reoccurring unintended events count is limited to Serious Safety Events. In addition, we need to distinguish between cases that fall into the same Mode Category but with different Mode fact patterns.

For example, medication errors may occur at the manufacturing, purchasing, storage, compounding, and administration steps. The circumstances of an error in labeling an IV bag are distinct from those related to the administration of the medication. Both types of error are linked in the successive steps of safe medication practices. However, an RCA action plan to prevent incorrect labeling of an IV bag will not prevent errors due to the other Latent Root causes at the point of medication administration.

For this reason, the definition of Repeat Events focuses both on the consistency of the facts (the Mode specifics) as well as the presence of the same Latent Root. To identify Repeat Events, it is necessary to ask a set of clarifying questions. Table 9.1 illustrates the questions and how the answers determine whether a subsequent unintended event should be counted as a Repeat Event.

Consider the following scenarios for multiple pressure injuries. The first example is of a pressure injury on the coccyx of a second patient caused by a lack of turning. This was related to inadequate patient assessment, in the same unit, and for the same reasons. This meets the definition of a Repeat Event.

In the same circumstances, but on a different unit, the case also meets the definition of a Repeat Event. There was a failure to spread the action plan from unit to unit within the facility. Likewise, in the same circumstances but in a different facility, the case again meets the definition of a Repeat Event. There was a failure to spread the action plan within the system.

In another scenario, if the second pressure injury occurs on the coccyx of a patient in the same unit but is related to skin breakdown with the dying process (a Kennedy terminal ulcer), the Latent Cause is different. This does not meet the definition of a Repeat Event.

Finally, consider the circumstance of a pressure injury related to the use of a nasal bridal, occurring on the nares of a patient. The device-related injury is a different set of facts than the injury on the coccyx. This does not meet the definition of a Repeat Event.

TABLE 9.1
Repeat Event Determination

Same Mode Specific Facts?*	Same Facility?	Same Department/ Unit?	Same Latent Root Cause Category?**	Repeat Event? Yes/No
Yes	Yes	Yes	Yes	Yes
Yes	Yes	Yes	No	No
Yes	Yes	No	Yes	Yes
Yes	No	Yes	Yes	Yes
Yes	No	No	Yes	Yes
Yes	No	No	No	No
No				No
*Same Physical Roots & Mode specifics			**Determined by RCA	

9.2 STRATEGIES FOR NO REPEAT EVENTS

To consider strategies for preventing Repeat Events, we look at strategies for the department in which the unintended event occurred, Mode Categories across the facility and organization, and systems reliability with the goal of zero harm.

9.2.1 REVIEW OF ACTION PLANS WITHIN THE DEPARTMENT/UNIT

The first strategy for no Repeat Events is to prevent the unintended event from reoccurring within the same department/unit (see Table 9.1). To measure progress toward achieving no Repeat Events, we inquire whether unintended events with the same or closely similar facts have previously occurred in the department/unit. If so, was the previous action plan not implemented or implemented but failed to prevent the same kind of harm event from reoccurring? These barriers must be understood and overcome since there should be no Repeat Events within the same department/unit after an RCA has been conducted.

9.2.2 REVIEW OF MODE CATEGORIES ACROSS THE ORGANIZATION

Next, we inquire whether an unintended event with the same Mode Category and Latent Roots occurred elsewhere within the facility or organization in a similar unit or department (see Table 9.1). If so, we need to know if the action plan was spread, if the spread of the action plan faced barriers, or if the spread failed to address adaptations needed for the local situation. Similarly, as within the unit, these barriers must be understood and overcome since there should be no Repeat Events after an RCA within the same facility or organization.

To aid in determining Mode Categories for focused improvement efforts across the organization, trending should be done at the Event, Mode Category, and Latent Root Cause levels.

9.2.2.1 Trending at the Event Level

At the Event level, the goal is to eliminate harm and have no Serious Safety Events. Monitoring progress toward this goal requires tracking and trending at the Event level. HPI[1] proposes a metric using the rolling 12-month average number of Serious Safety Events per 10,000 adjusted-patient-days. By using a rolling 12-month average, this measure tends to smooth out the data. That is, variation by month is not as visible as the overall direction and trend. However, the measure is quite good at indicating sustained improvement or worsening over time. If the trend line is improving, we can be quite confident that organizational change is occurring.

Another metric to consider is the ratio of RCAs done for Close Calls versus Serious Safety Events. Over time, as the rate of Serious Safety Events decreases, more resources for RCA may be devoted to the high-risk Close Call events. These are the gold to mine for preventing future harm.

9.2.2.2 Trending at the Mode Level

Mode Categories indicate the kinds of failures that are occurring. These are the triggering events for RCAs. Trending at the Mode Category level can help determine an improvement strategy. Think about the potential impact of implementing clinical initiatives, with standardization and the use of bundles, based on the Mode Category trends. Knowing the most frequent Mode of a Serious Safety Event can help determine whether to initiate improvement with pressure injuries, falls, infections, medication errors, diagnostic errors, or whatever is the greatest safety risk at the time. Improvement resources are valuable, and focusing the efforts at the highest priority of Mode Category is one way of prioritizing the work.

Initially, these may be common types such as medication errors, falls, pressure injuries, or diagnostic errors. Over time, as the spread of RCA action plans occurs, the most frequent Mode Categories may begin to shift. That is, improvement work can help reduce the most common types of Serious Safety Events. David found that the most frequent Modes in a comparison of three years of data showed shifts: whereas pressure injuries and patient falls had initially been among the top three, improvement work had reduced their number and when re-measured, they were no longer in the top three Mode categories.

Another factor to consider in trending Mode Categories is the inclusion of errors of omission. Using a definition for Serious Safety Events such as the one proposed by ASHRM[2], "a deviation from the generally accepted practice or process" is important for a comprehensive safety program. This involves the failure to act or to do something normally expected. These may involve failures to communicate, human error slips, lapses, low compliance, or gaps in training. Trending categories that denote errors of omission will help focus improvement work and clinical standardization beyond the individual RCA action plan.

Trends at the Mode Category level may also differ depending on the Event types included. The most frequent Mode Category for Serious Safety Events may differ from the most frequent Mode Category if Close Calls are included. For example, medication errors may rise to the top of the Mode Category trend if Close Call RCAs are included in the analysis. However, the rate of medication errors resulting in Serious Safety Events is relatively low. The category may not appear in the top if trending is limited to the Serious Safety Event RCAs. Both metrics are important and can help determine future strategies for improving safety.

9.2.2.3 Trending at the Latent Root Cause Level

Trending at the Latent Root Cause level allows a different view into the system vulnerabilities. These trends bring to light the holes in the Swiss cheese and help us predict how they might impact safety going forward. Knowledge of the most frequent types of Latent Root Causes provides insight into the strategies needed to improve the reliability and safety of the organization across Mode Categories.

Latent Roots are often described in a unique way by RCA teams. The language used by the RCA team reflects the specifics of the analysis and their understanding of the systems level issues. However, these idiosyncratic Latent Root descriptions can be categorized. The categorization allows trending. What emerges is a picture of distinct system-level vulnerabilities that could result in several different Modes.

Harold and David were asked to conduct a meta-analysis of the Latent Root Causes from all RCAs done during a specific period. The hospital administrator was interested in what might be learned about the impact of moving into a large, new facility.

The focus of this analysis was not on the Mode Categories, but on the system gaps and vulnerabilities: the Latent Roots. We have repeated this type of meta-analysis since that initial study, with remarkable insights. For example, repeated meta-analyses have found that problems with policies, procedures, or protocols, such as their absence, inadequacy, or confusing nature, result in numerous Modes of unintended events. Other examples of trending at the Latent Root level includes failures of communication among caregivers, problems with information availability, gaps in training, problems with human-equipment interfaces, and Human Factors.

Trending is done at each of these levels so that we know if Serious Safety Events are being reduced. We've found that trending at all three levels, Event, Mode, and Latent Root, is required to provide the information needed to strategically address Repeat Events.

9.2.2.4 Spreading Action Plans

In Chapter 5, we examined how to create an effective RCA action plan. We described the process for identifying a Latent Root to eliminate; targeting an action plan aimed at 100% elimination of the Latent Root cause; using outcome, process, and balancing measures to know whether the action plan is working as intended; and tools to support the implementation. The question we now address is how to spread these action plans beyond the local department or unit where an unintended event occurred and *proactively* implement the improvement to *prevent* the same unintended event from occurring elsewhere in the facility or organization.

From our experience, we recommend that different locations be considered as *unlike*. For example, an ICU in one hospital likely has different environmental factors, different culture, different equipment perhaps, from other ICUs even in the same system. Adaptation of RCA action plans, while adhering to the outcome targets, will be required. Take, for example, the spread of a falls-prevention action plan across a large number of clinics involving a standardized falls risk assessment. The clinics have several unique physical layouts. The workflows may vary for how patients are greeted, registered, and brought into exam rooms. The process measure of doing a falls risk assessment on each patient is aimed at achieving the outcome target of zero falls with harm. Each clinic must adapt the action plan to their workflow and environment.

While we spread and adapt RCA action plans, we also need to use prospective analyses. At the Mode Category level, we can use trending to help prioritize the highest candidates for RCA. Lean methods, such as copy-Kaizen (Yokotan) may be used to adapt and spread improvements. This is a method for sharing learning laterally across an organization, which entails copying and improving on improvement ideas that work. It is horizontal deployment, peer-to-peer, not top-down, with the expectation that people go see for themselves and learn how another area implemented their action plans and then improve on those ideas in the application to their area.[3]

The question of *how* to spread action plans will vary across healthcare organizations, depending on the size and structure. In our experience, there is a need in every healthcare organization for spreading all types of improvement. Safety improvement is one domain, but not the only one. In fact, using structures that include operations is critical. If the spread of action plans and safety improvement is in a silo apart from

TABLE 9.2
Repeat Event Strategies

Same Mode Specifics Facts?*	Same Facility?	Same Department/ Unit?	Same Latent Root Cause Category?**	Repeat Event? Yes/No	Strategy
Yes	Yes	Yes	Yes	Yes	PDCA Review Action Plan with Cause-Effect and Rigor Tests
Yes	Yes	Yes	No	No	New Action Plan
Yes	Yes	No	Yes	Yes	Adapt & Spread Action Plan
Yes	No	Yes	Yes	Yes	Adapt & Spread Action Plan
Yes	No	No	Yes	Yes	Adapt & Spread Action Plan
Yes	No	No	No	No	New Action Plan
No				No	New Action Plan
*Same Physical Roots & Mode specifics			**Determined by RCA		

the operations leadership, it is less likely to be acted upon. Many healthcare organizations have existing structures to leverage spreading action plans. These might include tiered huddles from the unit or department level up to the senior leadership. Some organizations have adopted a daily safety call or briefing. One multi-state medical group of primary and specialty care conducts a weekly review of all RCAs. The meeting is attended by the regional Medical Directors and Administrative Vice Presidents, along with support staff. The Chief Medical Officer facilitates the meeting in which a brief description of each unintended event, the findings of the RCA, and the action plan is shared. There is discussion on the potential risk for a repeat event across the medical group, and decisions for which areas should be targeted for spreading the action plans.

If a Repeat Event is confirmed, the strategies for preventing future reoccurrences vary. Table 9.2 suggests an appropriate strategy for each circumstance of Repeat Events.

Spreading RCA action plans to prevent Repeat Events also points to the need for clinical standardization. Such a large topic is beyond the scope of this book. We recommend checking resources such as the Healthcare Advisory Board white paper, "The System Blueprint for Clinical Standardization: Leveraging Systemness to Reduce Clinical Variation."[4]

9.2.3 SYSTEMS RELIABILITY

The third strategy is to spread the knowledge of reliable design gained from RCAs to build systems that support eliminating all unintended events and patient harm.

Trending at both the Mode Category and Latent Root levels facilitates the goal of a total systems approach to safety; working toward standardization of best practices associated with Mode Categories will help reduce those types of unintended events. But as the authors of *Free from Harm*[5] remind us, the goal of zero harm will not be achieved with project-based improvements. In other words, focusing improvement efforts only at individual Mode Categories will not achieve the goal of zero patient harm. Knowledge and action plans from individual RCAs must be compiled into design principles, adapted, and spread to address all the systemic factors that contribute to unintended events and patient harm. A procedural safety checklist is an example of a reliable system design. Checklists originated from specific unintended events but can be spread across all departments to help caregivers anticipate, mitigate, and eliminate other unrelated potential unintended events and patient harm.

9.3 SUMMARY

It is vital to understand that an RCA of an unintended event is conducted not just to understand and fix that specific event, but to prevent the unintended event from reoccurring. Multiple strategies are needed to achieve the target of no Repeat Events and zero harm. Trending of Events, Mode categories, and Latent Root causes helps anticipate harm events and prioritize improvement efforts. Spreading RCA action plans and safety improvement work reliably, and methods for achieving clinical standardization are needed. With a focus on eliminating Repeat Events, RCAs will increasingly be for "one-off" events.

NOTES

1. C. Throop and C. Stockmeier, *The HPI SEC & SSER Patient Safety Measurement System for Healthcare* (Virginia Beach, VA: Healthcare Performance Improvement, LLC, 2009).
2. M. Hoppes et al., *Serious Safety Events: A Focus on Harm Classification: Deviation in Care as Link Getting to Zero*. White Paper Serious-Edition No. 2 (Chicago, IL: ASHRM, 2012).
3. www.leanblog.org/2011/05/guest-post-what-is-yokoten/.
4. Health Care Advisory Board, "The System Blueprint for Clinical Standardization," 2015, https://preview.new.advisory.com/-/media/project/advisoryboard/shared/research/hcab/white-papers/2016/31775-hcab-the-system-blueprint-for-clinical-standardization.pdf?rev=c11ae67299304c1da718a110af9f93d1&hash=F8715E57AFC1F33205C11B14A081E971.
5. National Patient Safety Foundation, Free from Harm: Accelerating Patient Safety Improvement Fifteen Years *After to Err Is Human* (Boston, MA: National Patient Safety Foundation, 2015), 8.

10 Teaching RCA

It is our experience that to get closer to the goal of preventing reoccurring unintended events and eliminating harm, the knowledge, and practical skills for facilitating RCAs is essential. This book came about because of the lessons learned in leading and teaching RCA methodology in healthcare. Over the years, we have developed several training classes of various lengths. The most recent evolution resulted in an eight-hour training course. The contents of this book reflect the curriculum for that course. In this chapter, we provide the training outline and reflections on teaching. The reader may notice that the book is organized in the chronological order of the steps of an RCA. In contrast, when teaching RCA, we find it helpful to introduce the Logic Tree early and to continue to layer additional levels of understanding through the exercises and didactic material. The flow of the curriculum is based on our experience with adult learners. With the information contained in this book, combined with the training outline, we hope the reader will be enabled to be not only an effective RCA facilitator but also a teacher and champion of RCA.

Who should receive training on RCA? Certainly, those who are tasked by healthcare organizations to facilitate RCA teams. These can include quality leaders, risk managers, patient safety professionals, caregiver safety coordinators, and clinical leaders. The disciplines involved in RCA leadership vary across healthcare. Our belief is that the commitment to safety should not.

As outlined in Chapter 7, in addition to RCA facilitators, Executives, Directors, Managers, and others in leadership can benefit from training. We want them to understand how unintended events are prioritized and when to employ the tool of RCA. Executives and Process Owners play key roles in thorough and credible RCA and should be trained in the Logic Tree methodology and action planning. Other professionals who may serve as subject matter experts and RCA team members will be better prepared with advance training. Training for these roles may not require the full course. We encourage customizing the training to the audience.

In truth, the process of information gathering and using deductive reasoning to understand the origins of errors is beneficial beyond patient harm events. Asking the question, *"How could?"* can lead to deeper inquiry and problem-solving for all types of issues. In other words, there is no reason to limit RCA training to those responsible for follow-up on unintended events.

The following curriculum is based on our use of the Logic Tree methodology and the PROACT software. The curriculum can be adapted for paper-based RCA processes, as well as other software. The training can be delivered to groups or individuals. As in most learning environments, it is beneficial to have an optimal size group to allow full participation.

10.1 TRAINING PRE-WORK

In preparation for the RCA training, we want to confirm participants' registration, along with the dates, times, and locations. We ask participants to confirm they have loaded the software on their device and ask that they bring their computers to the training. Two articles are distributed for reading in advance of the training.[1] The selection of articles can vary. We have found team teaching to be helpful. In preparation for the training, the teaching team should review the curriculum and decide how the session will be facilitated.

Materials needed should be confirmed in advance. Table 10.1 is a list of materials we prepare. Packets for participants are made up in advance. We include the training guide, copies of the tools used, and the training exercises.

10.2 TRAINING GUIDE

The following training guide includes approximate times for each section, an outline of the main points covered, and teaching notes. We have broken the eight-hour training into two sections: our experience is that participants are better able to absorb the material. Typically, the first four-hour session is held in an afternoon, followed on the next day by a four-hour morning session.

TABLE 10.1
Training Materials Checklist

Pre-Work	☐ Confirm software availability for participants
	☐ Read: Robert J. Latino, Root Cause Analysis versus Shallow Cause Analysis: What's the Difference? (available at archive.reliability.com/pdf/Root-Cause-Analysis-vs-Shallow-Cause-Analysis-ENG2.pdf)
	☐ Read: Robert J. Latino, Improving Reliability with Root Cause Analysis. September/October 2013. Patient Safety & Quality Healthcare (PSQH) (available at psqh.com/analysis/improving-reliability-with-root-cause-analysis/)
Training Materials Needed	☐ Software
	☐ Prioritization Tool
	☐ Hierarchy of Analytical Tools
	☐ Logic Tree Definitions diagram
	☐ Swiss cheese model
	☐ Action Plan template
	☐ Training Exercises: Leader's Guide
	☐ Exercise #1
	☐ Exercise #2
	☐ Exercise #3
	☐ Root Cause Analysis Training Guide for each participant
	☐ Post-it Notes
	☐ Easel Pad(s)

10.2.1 Training Outline

10.2.1.1 Day 1
Welcome and Reflection (10 minutes)

- Do introductions if needed
- Provide a reflection on the value of RCA and your organization's commitment to safety

Review Objectives and Agenda (20 minutes)

- Purpose of training
  - Increase understanding of RCA
  - Increase capacity for RCA facilitation
- Objectives
  - Understand when to apply RCA and other analytical tools
  - Demonstrate knowledge of Physical, Human, and Latent Root Causes
  - Demonstrate ability to construct a Logic Tree evidencing cause and effect
  - Demonstrate knowledge of the software
  - Demonstrate knowledge of action planning strategies
- Agenda Review
  - *Review the entire agenda.*
- Participants' Goals
  - *Invite participants to identify their learning goals; pay attention to what role in RCA they will play. Say what the training does not cover: FMEA, software proficiency.*

The Need for RCA (30 minutes)

- Organizational reasons
  - *Discuss the reasons for conducting RCA: high reliability, organizational learning, safety events, operational/quality/regulatory needs.*
- Events and repeat events
  - *Discuss the appropriate use of RCA for individual events and for repeated failures.*
- Other forms of analysis
  - *Describe and discuss the use of Troubleshooting; Brain Storming; 5 Whys; Case Review; Peer Review; Failure Modes and Effects Analysis (FMEA); Meta-Analysis. Share the Prioritization Tool and Hierarchy of analytical tools.*

Introduction to Preparing an RCA (75 minutes)
Confirm the correct software version and access (if applicable).

- Describe steps in the RCA process
  - *Introduce the acronym PROACT: Preserve information; Order the team; Analyze the event; Communicate results; Track action plans.*
- Define the focus of RCA
  - Event Types
    - *Introduce the categories of Events: Serious Safety Events, Close Calls, Operational Events, Quality/Performance Improvement.*
  - Repeat Events
    - *Discuss types of repeat events and reasons for RCA.*
  - Modes
    - *Discuss the definitions for Modes & their basis in policy, and best practices (NQF, ASHRM, TJC, state rules & regulations).*
- Introduce the Top Box
  - Define Event, Mode, Multiple Modes
  - *Draw a Top Box; discuss Event and Mode definitions; discuss when to use multiple Modes; introduce concepts of failure and response to failure as two possible Modes; draw a line around the Top Box; stress the importance of the Mode being factual and accurate.*
- Introduce Hypothesizing
- Use of "How could?" versus "Why?"
  - *Discuss the value of the question, "How could?" versus "Why?"*
- Two forms of initial hypotheses
  - *Introduce the use of deductive reasoning and large buckets for initial hypotheses; give examples; introduce the use of process steps as hypotheses.*
- Continue hypothesizing
  - *Demonstrate hypothesizing down the Logic Tree.*
- Risks of hypothesizing
  - *Describe risks of hypothesizing; practice initial hypothesizing for common Modes.*
- Introduce Types of Root Causes
  - Physical Roots
  - Human Roots
  - Errors of commission and omission
  - Human error versus Human factors
  - Latent Roots
  - Contributing factors
  - *Use the Logic Tree Definitions diagram for review and discussion of the definitions; note how initial hypotheses are Physical Roots; introduce the concept of two types of Human Roots – errors of commission and omission; describe James Reason's concepts of Human error; discuss Human Factors as a Latent Root category; define Latent Roots and Contributing Factors.*
  - Cause-and-Effect Chain
    - *Demonstrate a test of the chain of causation using a cause and effect path on the Logic Tree, showing a pathway to failure.*

Exercise #1: Logic Tree (45 minutes)

- Develop a Logic Tree using the chronology and data provided
  - *Provide Exercise #1; review the directions; invite participants to work independently; circulate and provide feedback.*
- Share Logic Trees with the group
  - *Invite the group to share their Logic Trees, including Top Box, Physical Roots, Human Roots, Latent Roots, Contributing Factors, Chain of Causation, Verifications; note the similarities of Latent Roots.*

Pre-work of the RCA Team Meeting (30 minutes)

- Preserving data
  - Types of data
  - *Discuss the types and sources of data; discuss sequestering.*
  - Interview techniques
  - *Discuss interview strategies and quick tips.*
- Ordering the team
  - *Discuss RCA team composition.*
- Building the Top Box
  - *Describe the practice for building Top Box prior to team meeting.*
- Chronologies
  - *Discuss the purpose of chronologies; the scope, types, and sources of information.*

Exercise #2: Logic Tree (30 minutes)

- Develop a Logic Tree using the chronology and data provided
  - *Provide Exercise #2; review the directions; invite participants to work in pairs. There will be additional time to complete the exercise on Day 2.*
- Questions/Discussion

Day 1 Wrap-Up

10.2.1.2 Day 2
Reflection (10 minutes)

Day 1 Review (20 minutes)

- Questions and review from Day 1
  - *Briefly review the topics covered on Day 1; invite questions and observations.*

View YouTube Colombini MRI Case (60 minutes)

- Colombini MRI Case: Root Cause Analysis: Tobias Gilk & Reliability Center (www.youtube.com/watch?v=0nA-UceHMqc&t=1335s)

- *Introduce the YouTube video.*
- Group discussion
  - *Invite observations; discuss the use of the floor plan, diagram, and animation with the chronology; compare and contrast the analysis to the use of 5 Whys and Fishbone methodologies.*

Exercise #2 (Continued) (30 minutes)

- Continue the Logic Tree construction
- Share Logic Trees with the group
  - *Invite observations on the exercise; discuss the Mode; review Latent Roots and how they are different than Contributing Factors; answer questions.*
- Group Discussion: Latent Roots and Contributing Factors, Boolean Logic, Path to Failure with Chain of Causation, Verifications, Confidence Levels
  - *Using the Logic Tree from Exercise 2, introduce conjunctive and disjunctive Boolean logic; demonstrate the path to failure using the chain of causation; introduce verifications and confidence levels; answer questions.*

Action Planning (15 minutes)

- Review the concepts of Swiss cheese and reliability; discuss facilitation of choosing a Latent Root for action planning
  - *Using the action plan template, introduce the action plan format; demonstrate the selection of one Latent Root for action planning using the Swiss cheese model.*
- Introduce the rigor test, metrics, and improvement tools
  - *Introduce the rigor test; define and discuss the three types of metrics; review the improvement tools to support the action plan.*

Exercise #3: Repeat Failures (60 minutes)

- Divide into groups of two or three
  - *Working in small groups, one person provides facts for a case of repeat failures (which they come up with), a second person facilitates asking "How could?" and creates a Logic Tree, a third person can serve as team member/observer and provide feedback; switch roles after 15 minutes and repeat the exercise (note it is not necessary to complete a Logic Tree); this exercise helps develop the skill of constructing a Logic Tree in real time as information is gathered.*
- Debrief
  - *Invite participants to share their experience as the person being interviewed, the team member/observer, and facilitator.*

RCA Facilitation (30 minutes)

- Software and paper versions
  - *Discuss decisions about using paper and software-based RCA.*

- Team facilitation approaches and roles
  - *Encourage team facilitation; discuss the division of responsibilities; suggest best facilitation practices.*
- Skill development
  - *Suggest ways to practice Logic Tree construction and reviews.*

Wrap-Up Day 2 (15 minutes)

- Next steps for continued learning and practice
  - *Invite participants to share their commitments for next steps.*
- Feedback
  - *Record positives and recommendations from the training.*

10.2.2 Teacher's Guide

Instructions for the three training exercises follow. Exercises 1 and 2 include the following:

- A description of the Event for use in constructing the Logic Tree.
- Directions for the exercise.
- A Timeline meant to be illustrative more than useful in the training. Two slightly different styles suggest to students that the specific form of the flow diagram, chronology, or sequence of events can vary.
- Logic Trees as examples. There may be variation in the trees constructed by students in the training. Look for the commonality of the Latent Roots identified by students as validation of the methodology, evidenced by different investigators with different looking trees arriving at the same Latent Roots.

10.2.2.1 Training Exercise #1

This exercise uses a non-clinical example. The assumption is that it may be easier to see the logic in a less complicated setting than healthcare. One goal of this exercise is to reinforce the understanding of the types of roots: Physical, Human, and Latent. A second goal is to develop the deductive reasoning inherent in the Logic Tree. After students have completed the exercise, the teacher may demonstrate confirming the cause-and-effect chain from the Latent Roots up to the Mode. The teacher may also note the confirmed hypotheses at the top level are Physical Roots.

10.2.2.2 Training Exercise #2

This exercise is more complex. The teacher may want to lead a discussion about how to define the Mode in this case. The students are prompted in the exercise to think about what is required to successfully complete the surgery. The teacher may want to talk about this as a strategy for identifying hypotheses. The discussion can include the ways in which the hypotheses should be framed as failure statements. Students should be able to find roots for three hypotheses: the lack of information about the specialty mesh need, the lack of stock, and the failure to know about the problem prior to inducing anesthesia.

10.2.2.3 Training Exercise #3

The goal of this exercise is to increase proficiency in asking the *"How could?"* questions while concurrently constructing a Logic Tree. Invite the students to work in groups of two or three. One student will construct a Logic Tree and gather the information needed by asking the partner(s) about hypotheses and evidence. A second student should think of a repeated failure to provide the hypotheses and information for the Logic Tree. (Examples are repeat triage failures, repeat scheduling errors, repeat staffing problems, or repeat medication reconciliation errors.) If there is a third student, they may participate as a team member and contribute to the hypotheses.

Allow teams to work for approximately 15 minutes and then switch roles. Allow sufficient time to debrief as a class. Invite students to share their experiences: how did it feel to be the investigator, the subject matter expert, and the third person. What were their observations about the process? Discuss the implications of this awareness for their RCA team facilitation.

10.2.2.4 Student Handout Training Exercise #1: Train Derailment

You have been assigned to investigate and conduct an RCA after a passenger train derailed in the mountains between Chemult and Oakridge, Oregon. Miraculously, no passengers were hurt. The engine was a total loss. Interviews and data collection have revealed several facts. To begin, label the following as Physical, Human, or Latent roots:

- The most common cause of a derailment is problems with the track. An engineer has examined the track and did not find any breaks.
- The accident occurred on a curve, as the train was coming down the mountain. The engine was found lying on its side on the outside of the curve.
- The curve is marked at 45 mph.
- An examination of the "black box" reveals the speed registered deceleration from 70 mph to 50 mph prior to the accident.
- The conductor tells you the train was behind schedule that day.
- The train is often behind schedule because freight trains do not yield the right of way to the track.
- Examination of the brakes reveals less than 10% braking capacity remaining.
- A maintenance supervisor tells you this engine was three months overdue for routine maintenance.
- Funding cuts have caused re-prioritization of the maintenance schedule.

10.2.2.5 Student Handout Training Exercise #2: Canceled Surgery

Your team has been assigned to investigate and conduct an RCA after the cancellation of a surgery due to the lack of a necessary supply. A special mesh, preferred by the surgeon for certain difficult abdominal repairs, was not available in the operating room. Unfortunately, the patient had already been sedated. The case was canceled and rescheduled.

DIRECTIONS

Additional information has been obtained. Using the following facts, create a Logic Tree.

- Label the Event: "Training Exercise #2."
- Label a Mode based on the condition that to successfully start the surgery, the special mesh should have been in the operating room *and* supplies should have been confirmed before anesthesia was induced.
- Review the results of the interviews and information gathered by your team. Label each item as a Physical root, Human root, or Latent root.
- Construct a Logic Tree.

Facts:

- Ms. Fortune initially saw her surgeon, Dr. Dogood, in the clinic. Dr. Dogood recommended surgery to revise an abdominal repair that was leaking and discussed possible risks and benefits of the procedure. Ms. Fortune agreed and the procedure was scheduled for the next week.
- Dr. Dogood asked her staff to complete the consent form and to schedule the surgery. The office nurse filled out the consent form for Dr. Dogood to have Ms. Fortune sign.
- The office clerk faxed a copy of the request for surgery time, along with the consent form, to the hospital scheduler.
- When reviewed, the surgery schedule does not indicate any special equipment or supplies are needed.
- The forms received from the surgeon's office were not kept. However, the scheduler tells you that if they included any information about special mesh, this would be noted on the hospital surgery schedule.
- The scheduler adds Ms. Fortune to the surgery schedule.
- The surgery charge nurse reviews cases for the next day, including Ms. Fortune's.
- The surgery charge nurse tells you there are occasionally errors on the surgery schedule when information is missing or wrong. She doesn't know what happened in this case.
- The surgery tech tells you that when it was discovered the special mesh was not on the back table, he called to the supply room. He was told they had one special mesh but that it was outdated and could not be used.
- The surgeon states the special mesh is only used in complicated re-operations. These cases are rare.
- The sterile processing manager tells you the special mesh is only used once or twice a year. Her records show two were purchased last year, and one was used.
- Sterile processing does not have a process for checking outdates on specialty supplies. The practice has been to re-order when the supply is used.

- The day of surgery, the circulating nurse brings Ms. Fortune into the Operating Room and confirms her name and procedure.
- Ms. Fortune is sedated by the anesthesiologist.
- The surgeon arrives at the room and leads a final time out. She asks whether the special mesh is available.
- The team realizes the supply is not available and contacts Sterile Processing.
- The case is canceled and rescheduled.

NOTE

1. Robert Latino, "Root Cause Analysis versus Shallow Cause Analysis: What's the Difference?" www.archive.reliability.com/pdf/Root-Cause-Analysis-vs-Shallow-Cause-Analysis-ENG2.pdf; Robert Latino, "Improving Reliability with Root Cause Analysis," *Patient Safety & Quality Healthcare (PSQH)*, September–October 2013, www.psqh.com/analysis/improving-reliability-with-root-cause-analysis/.

References

American Society for Healthcare Risk Management (ASHRM). *Serious Safety Events: A Focus on Harm Classification: Deviation in Care as Link Getting to Zero.* White Paper Serious-Edition No. 2. Chicago, IL: ASHRM, 2014.

Clapper, C., J. Merlino, and C. Stockmeier. *Zero Harm: How to Achieve Patient and Workforce Safety in Healthcare.* New York: McGraw-Hill, 2019.

Gordon, S. *Beyond the Checklist.* Ithaca and London: ILR Press, 2013.

The Joint Commission. *Root Cause Analysis in Healthcare: A Joint Commission Guide to Analysis and Corrective Action of Sentinel and Adverse Events,* 7th ed. Oakbrook Terrace, IL: Joint Commission Resources, 2020.

———. "Sentinel Events." www.jointcommission.org/resources/patient-safety-topics/sentinel-event/.

Latino, M., R. Latino, and K. Latino. *Root Cause Analysis: Improving Performance for Bottom-Line Results.* Boca Raton: CRC Press, Taylor & Francis Group, 2020.

Latino, R. *Patient Safety: The PROACT® Root Cause Analysis Approach.* Boca Raton, FL: CRC Press, Taylor & Francis, 2009.

Liker, B. *Becoming Lean.* Portland: Productivity Press, 1998.

MacLean, N. *Young Men and Fire.* Chicago, IL: University of Chicago Press, 1972.

National Patient Safety Foundation. *RCA2: Improving Root Cause Analyses and Actions to Prevent Harm.* Boston, MA: National Patient Safety Foundation, 2015.

———. Free from Harm: Accelerating Patient Safety Improvement Fifteen Years after *To Err Is Human.* Boston, MA: National Patient Safety Foundation, 2015.

National Quality Forum (NQF). *Serious Reportable Events in Healthcare—2011 Update: A Consensus Report.* Washington, DC: NQF, 2011.

Reason, J. *Human Error.* New York: Cambridge University Press, 1990.

———. "Human Error: Models and Management." *British Medical Journal* 320, no. 7237 (March 18, 2000): 768–70.

Schein, E. *Humble Inquiry: The Art of Asking Instead of Telling.* Oakland, CA: Berrett-Koehler Publishers, 2013.

Syed, M. *Black Box Thinking.* London: John Murray, 2015.

Throop, C., and C. Stockmeier. *The HPI SEC & SSER Patient Safety Measurement System for Healthcare.* Virginia Beach, VA: Healthcare Performance Improvement, LLC, 2009.

U.S. Institute of Medicine. *To Err Is Human: Building a Safer Health System.* Washington, DC: National Academies Press, 1999.

Appendix 1
Logic Tree for Train Derailment Exercise

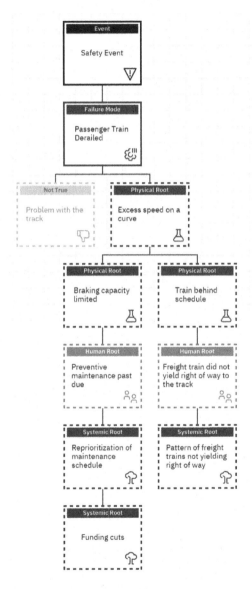

FIGURE APPENDIX 1: Logic Tree for Train Derailment

Source: EasyRCA image used by permission from Reliability Center, Inc.

Appendix 2
Logic Tree for Canceled Surgery Exercise

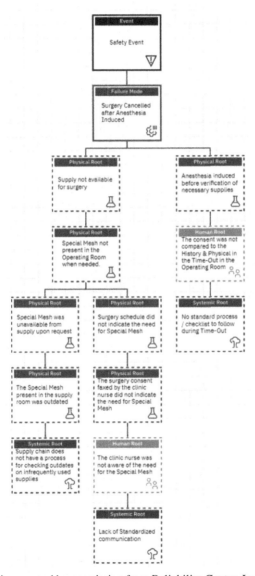

Source: EasyRCA image used by permission from Reliability Center, Inc.

Index

Note: Page numbers in *italics* indicate a figure and page numbers in **bold** indicate a table on the corresponding page.

5 Whys, 12–13, 19–20, 42–43

A

A3, 12
abnormality tracker, 69–70, **70**
action plans
    abnormality tracker, 69–70, **70**
    elements of, 64
    metrics, 67
    novel approach to, 61–64
    process observation, 69, **70**
    reality of, 61
    review of, 101
    rigor testing, 65, **65**
    rounding, 70
    spreading of, 103–104, **104**
    standardized work, 67–68
    strength levels, *62*
    template, 65–67, **66**
    training matrix, 69, **69**
active facilitation, 76–77
additional recommendations page, 76
analysis, *see also* Logic Tree
    5 Whys, 12–13
    A3 thinking, 12
    apparent-cause analysis, 14
    case quality review, 14
    complex problem-solving, 12
    Failure Modes and Effects Analysis (FMEA), 13–14
    meta-analysis, 15
    morbidity and mortality (M&M) conferences, 16
    opportunity analysis (OA), 14
    peer review, 15
    selection of form of, 16–18
    simple problem-solving, 11–12
    troubleshooting, 11
apparent-cause analysis, 14

B

balancing measure, 67
barriers, to RCA
    countermeasures, 92–96
    reasons for RCA failures, 89–92
bimodal methods, 20
Boolean logic, 44, *44*, *45*, 59

C

caregivers
    caring for, 3–4
    identification of, 27
    as source of evidence, 50–51
    standardized work, 84–85
case quality review, 14
case studies, for RCA training, 86
categories, *see* mode categories
causal chain, 78
causation, chain of, 57, *58*
checklist
    for evaluation, 94–95
    for RCA team membership, **28**
    for training materials, **108**
chronology, 30–31, 35–36, 76–77
clinical leaders, 27, 29, 30, 33
Close Calls, 16, 27, 31, 39, 99
Closure Huddle, 67
compassion and empathy, 3
complex problem-solving, 12
confidence, 55–57
confidentiality, 29–30
conjunctive Boolean logic, 44, *45*
contributing factors
    definition of, 59
    versus latent roots, 59
countermeasures, 92–93
    for Action Plans were not effective, **93**
    for Action Plans were not implemented, **93**
    for broad hypotheses, 46–47
    for categories versus hypotheses, 45–46
    and Executive Sponsors, 94
    for latent roots as hypothesis, 46
    for Latent Roots were not found, **93**
    for narrow hypotheses, 47–48
    and organization, 94
    RCA evaluation template, **95–96**
    RCA evaluation tool, 94–95
    and RCA facilitation, 93–94
Crew Resource Management (CRM), 7

D

deductive reasoning, 14, 47, 59, 78, 107
disjunctive Boolean logic, 44, *44*

Printed in the United States
by Baker & Taylor Publisher Services